Mnemonics in Ophthal

Fourth Edition

Edsel B. Ing, MD, FRCS (C)

Illustrated by Sabrina Ing, MD

Edited by Justin E. Anderson, MD

APOLLO™ AUDIOBOOKS

LUBBOCK, TEXAS

Find this and other titles available in

Digital Audio, Text Download, and Print at

www.ApolloAudiobooks.com

Mnemonics in Ophthalmology, Fourth Edition

DISCLAIMER

This book is intended to provide accurate information in accord with accepted medical standards at the time of publication. However, in a time of rapid change, it is difficult to ensure that all medical treatment and drug therapy information provided is entirely accurate and up-to-date. Despite the author's and publisher's efforts, in view of the possibility of human error or changes in medical science, neither the author nor the publisher nor any other party who has been involved in the preparation or publication of this work warrants that the information contained herein is in every respect accurate or complete, and they disclaim any liability, loss or risk, personal, financial or otherwise, for any errors or omissions or for the results obtained from the use of the information contained in this work. Readers should use this book only as a general guide and are strongly encouraged to confirm information contained herein with other sources.

Request for permissions to reproduce any portion of this book should be addressed to
Apollo Audiobooks, LLC
ATTN: Permissions
3204 30th Street
Lubbock, Texas 79410

Apollo Audiobooks is a registered trademark of Apollo Audiobooks, LLC
Lubbock, Texas
www.ApolloAudiobooks.com

Illustrated by Sabrina Ing, MD
Edited by Justin E. Anderson, MD

DEDICATION

To Helen, Mercedes, Royce & Max, for time lost,
Mom & Dad for their constant encouragement,
And to the residents and fellows, who must endure.

Edsel B. Ing, MD, FRCS(C)

Special Thanks from the Publisher

To Dean Xu, MD, MS and Hari Bodhireddy, MS IV
Of Texas Tech University Health Sciences Center, Lubbock, Texas:
Whose tireless efforts in fully updating the text and references and obtaining Permissions for
many of the mnemonics contained herein has made this updated
Fourth Edition of Mnemonics in Ophthalmology possible.

INTRODUCTION

The more you read, the more you learn.
The more you learn, the more you forget.
The more you forget, the less you know.
So why study?

Origin unknown

Mnemonics are no substitute for logic and clinical experience. However they are stepping stones to initiate thought, and reminders of things we might otherwise overlook.

This book should benefit ophthalmology residents, ophthalmic practitioners, and those involved in medical education. The mnemonics are based on letter associations, word associations, rhymes, and visual imagery techniques. Some of the mnemonics dea with core ophthalmic knowledge. Others are directed at the "ophthalmic trivia" we encounter at rounds or on exams. Where possible I have endeavored to make the mnemonic stem relate to the diagnosis. For example, the differential for "salt and pepper" retinopathy incorporates the word "COOKS."

Most of these mnemonics are of my own invention. I have attempted to obtain written permission and give appropriate credit to the mnemonics which are not my own. Several popular mnemonics that are ubiquitous and simply passed down from resident to resident are referenced as "Mnemonic origin unknown."

I hope these mnemonics will accelerate your office practice and add breadth to your differential diagnosis lists. For those in training, the memory techniques may afford you more free time, better sleep before exams, and less squirming during rounds. Perhaps it will also benefit individuals like me who will have to recertify for board exams each decade. Many areas remain uncovered in this short text. As I become more immersed in my subspecialties it becomes difficult to maintain the mnemonics in the other subspecialties. I will try to update the book regularly, and I do appreciate any comments, mnemonics contributions, or corrections.

Edsel Ing, 2009

Edsel Ing, MD, FRCS(C) is Assistant Professor at the University of Toronto.
He may be contacted at: Toronto Eyelid, Strabismus & Orbit Surgery Clinic, Toronto East General Hospital, K306, 650 Sammon Ave., Toronto, ON, M4C 5M5, Canada.
Phone: 416.465.7900, Fax: 416.385.3880

ACKNOWLEDGEMENTS

I much appreciate the opportunity to work with Justin E. Anderson, MD and Apollo Audiobooks. I thank Sabrina Ing and Mercedes Ing for their efforts, their artwork, and their tolerance of my requests for revisions. I greatly appreciate the enthusiasm, suggestions, and mnemonics of Dr. Mike Myles, and the contributions from the following physicians:

Gary Brown
Ray Buncic
R. Jean Campbell
Dave Chacko
Roger Dailey
Joe Flanagan
Elliot Finklestein
Robert Folberg
E. B. Glick
Stephen Goldberg
R. Bruce Grene
Pierre Guibor
Brent P. Hazen
J. W. Henderson
D. R. Johns
Anthony Johnson
Steve Kraft
Ravi Krishnan
G. B. Krohel
Jacqueline A. Leavitt
N. S. Levy
Richard Lindstrom

Brent MacInnis
Jeff Martow
John A. McWhae
Mary Mehaffey
Jose Ochoa
Robert O'Connor
John A. Parker
Bob Pashby
M. S. Ramsey
Mike Rauser
Esther Rettig
John J. Shipman
Michael Slavin
J. Lawton Smith
Rosa Tang
Dai Tran
Graham Trope
George O. Waring III
Joe Weinstock
Eliott Werner
William Willis
Aaron Zuckerberg

TABLE OF CONTENTS

I. USING MNEMONICS IN OPHTHALMOLOGY

MNEMONICS IN OPHTHALMOLOGY

Let's begin with some quick points on how to use mnemonics in your ophthalmology practice and on ophthalmology exams.

1) Mnemonics are no substitute for common sense. Comprehension should always accompany memorization. When giving a differential diagnosis, state the most likely answer first. THEN use the mnemonic to fill in the gaps.

2) No book can provide a mnemonic for everything. However the examples in this text demonstrate the different types of mnemonics that can be employed in ophthalmology. Since we each organize information uniquely, some of the best mnemonics will be self-contrived. A mixture of "right brain and left brain" techniques is helpful. Endeavor to make the stem word relate to the actual diagnosis, since this will enhance long-term retention. An ideal mnemonic will incorporate more than one of the following techniques: letter association, word association, rhyme, or visual imagery.

Further references for the interested reader include:

i) Patten BM, "The history of memory arts," Neurology: 1990:40; pp. 346-352.

ii) von Oech R, A Whack on the Side of the Head, 25th Anv Rev Ed. Business Plus Pubs, 2008.

iii) Lorayne H, Lucas J, The Memory Book. Ballantine Books, 2000.

3) When faced with a difficult problem interface logic with mnemonics. It is often useful to approach such problems from a patho-anatomic perspective asking "Where the lesion is, and what the lesion is?" Let's take the example of unexplained visual loss. The problem may be anywhere from the refractive correction, to the tear film, to the occipital lobe, or be supratentorial. Write down each of these "anatomic" areas in a vertical column at the left hand of the page. Each anatomic area can be affected by various diseases, which you then list in a horizontal row at the top of the page. Then fill in the blanks of your table. The mnemonic I use for a general DIFFERENTIAL DIAGNOSIS IN OPHTHALMOLOGY is:

CV INDICATES MY HI'R CLASS

General	*Focused*
Congenital	**MY**asthenia, **M**asquerade syndromes
Vascular	**HI**V, **H**ereditary
Inflammatory, **I**nfectious	**R**efractive
Neoplastic	
Degenerative, **D**ystrophies, **D**emyelinating	
Iatrogenic (including medications)	**C**arcinomatous meningitis
Collagen vascular	**L**yme disease
Allergy	**A**rteritis (GCA), **A**myloid, **A**mblyopia
Trauma	**S**arcoid
Endocrine (incl. metabolic & pregnancy)	**S**yphilis
S	

II. OPTICS

OPTIMAL PINHOLE APERTURE

Pinhole occluders can improve the focus of light in the ametropic eye by reducing the size of the blur circle. There is an "optimal" size for the pinhole aperture which can be remembered as follows:

The **1, 2, 3** Rule for the optimal pinhole aperture.
A **1.2** mm pinhole corrects up to **3** D of refractive error.

Information source: James Vander & Janice Gault (eds), Ophthalmology Secrets. Mosby, 2007, p. 29.

GLASSES PROBLEMS

If a patient is unhappy with his or her glasses, the following things should be checked. Remember the causes for patient dissatisfaction with spectacles with this little rhyme:

Base curve, center, Pow/Ver, Phoria, Seg Height, Plus Tilt and **Arm length check** if the lens isn't right.

Base curve
Center (Optical center, Facial center i.e. Facial asymmetry)
Pow (Lens power)
Ver (Vertex distance)
Phoria
Seg Height (Bifocal segment height)
Tilt (Pantoscopic tilt)
Arm length check (Reading distance preferred by the patient)

EFFECTIVE POWER OF A LENS

The effective power of lens is affected by its distance from the eye. You can gain effective plus power by increasing the vertex distance from the eye.

You can "**+**" (increase) the power of a lens by "**+ing**" (increasing) the vertex distance.

Thus, under-corrected presbyopes wear their half glasses as far down their nose as possible.

Mnemonic by permission of and modified from John A. Parker, MD, FRCS(C).

TYPES OF HYPEROPIA

There are several types of hyperopia that contribute to the patient's overall refractive error. These include total, latent, manifest, absolute, and facultative hyperopia.

Remember "**TeLMeAF**" for the types of hyperopia.

"**TeLMeAF**" sounds like "Tell Me Off."
And the small "**e**'s" are "**e**"qual signs (=).

Total hyperopia = **L**atent + **M**anifest
Manifest hyperopia = **A**bsolute + **F**acultative

AVERAGE ACCOMMODATION IN THE MIDDLE-AGED	As patients lose accommodation and start becoming presbyopic in their 40's, you can estimate their remaining accommodation with the following rule:

mnemonic

The Rule of 4's for estimating accommodation in presbyopes.

Accommodation at age **44** is ≈ **4.4 D**	(4.5D+/-1.5D)
4 years **younger, add 1.4 D** i.e. age 40 is ≈ 5.8 D	(6.0D+/-2.0D)
4 years **older, subtract 1.4 D** i.e. age 48 is ≈ 3.0 D	(3.0D+/-1.5D)

Note: The bracketed () numbers are actual figures from the BCSC text.

Information source: American Academy of Ophthalmology Basic and Clinical Science Course: Clinical Optics. American Academy of Ophthalmology, 2008, p. 148.

THE ULTEX BIFOCAL

Ultex bifocals are unique because they, unlike other bifocals, are a plus cylinder grind. The cylinder is ground on the front surface of the lens since the add is ground on the posterior surface.

mnemonic

Ulte**X** bifocals are ground in **+** cylinder with the astigmatism correction ground on the conve**X** surface.

When you see the letter "**X**" in the word Ulte-"**X**," simply rotate it 45 degrees. The rotated "**X**" now forms a "**+**" sign, reminding you that Ulte**X** bifocals are ground in **+** cylinder with the astigmatism correction ground on the conve**X** surface.

KNAPP'S RULE & ANISEIKONIA

In eyes with axial ametropia, if the corrective lens is placed at the anterior focal point of the eye, no image magnification will occur. The anterior focal point of a typical eye is 17 mm.

mnemonic

Use the letters in **KNAPP**'s rule in **Ax**ial ametropia to remember the rule. (See Illustration)

Remember that when you **NAP**, you can sleep with an **Ax**e for protection from the letter **K**.
The **Ax**e reminds you that this rule applies to eyes with **Ax**ial ametropia.
The **AP** reminds you that the corrective lens should be placed at the **A**nterior focal **P**oint.
If the **Ax**e splits the letter **K** at its point of intersection, the mirror image of the second part of the letter is a lopsided 7 (**>**). Together the 1 and **>** form 17 mm, the typical anterior focal point for the eye.

THE DUOCHROME TEST IN REFRACTION

The duochrome test in refraction can help you refine a patient's spherical power in the phoropter. If the patient sees red side clearer add minus. If the patient sees green side clearer add plus.

mnemonic

Remember the **RAM GAP** rule for the duochrome test in refraction.

If the **R**ed side is clearer **A**dd **M**inus.
If the **G**reen side clearer **A**dd **P**lus.

Mnemonic origin unknown. I have heard it several times during my training.

THE DUOCHROME TEST FOR MALINGERING

There are many tests that can be used to help determine if a patient is malingering or has "functional vision loss." One of these is the duochrome test.

Remember – "**Green over Good**" with the duochrome test for malingering.

Put the **Green** lens **over** the **Good** eye when testing for malingering since the green lens is a more effective filter

Information source: EB Ing, Functional Visual Loss and the Duochronme Test, Can J Ophthalmol, 1995; 30: pp. 35-36.

TYPES OF LENS ABERRATIONS

Sadly, lenses do not obey first order optics. The thicker the lenses, the more aberrations will occur. The different lens aberrations can be recalled by the "**SAD 3C's**" story.

The "**SAD 3C's**" story of lens aberrations. (See Illustration)

Picture the **SAD** scene of a man in a hospital bed with **very thick glasses** (glasses subject to lens aberrations) who is **Comatose** after a car accident, with a **Chrome Curvature** (old style car fender) sticking out of his chest.

Spherical aberration
Astigmatism of oblique incidence
Distortion

3C's
Comatose (**C**oma)
Chrome (**C**hroma)
Curvature (**C**urvature of field)

TYPES OF DISTORTION AND REFRACTIVE ERROR

High power lenses will result in distortion of the image. High plus lenses create pincushion distortion while high minus lenses create barrel distortion.

mnemonic

Remember - high **P**lus lenses result in **P**incushion distortion.

ACTION OF POLARIZED LENSES

Unwanted glare usually results from scattered horizontal rays of light. These horizontal rays are preferentially filtered by polarizing lenses in glasses.

mnemonic

Think - **H**orizontal rays are **H**alted & **V**ertical rays ad**V**ance with polarized lenses in glasses.

THE ABBE NUMBER AND CHROMATIC DISPERSION

The chromatic dispersion of a lens can be expressed in terms of its Abbe number. A higher Abbe number means less chromatic dispersion, which is preferred.

mnemonic

Think of a group of competing church abbeys, each with a church steeple, and stained glass windows (representing chromatic dispersion). The church with the highest abbey steeple (meaning the highest Abbe number) is "the best" (with the least chromatic dispersion). (See Illustration)

Information source: American Academy of Ophthalmology Basic and Clinical Science Course: Clinical Optics. American Academy of Ophthalmology, 2008, p. 50.

GOLDMANN TONOMETRY WITH HIGH ASTIGMATISM

For every 4 diopters of astigmatism, the maximum error in Goldmann applanation tonometry is 1 mm Hg. To minimize this error in patients with high astigmatism, the red line on the prism should be aligned with the minus axis of the patient's corneal astigmatism using the lines on the sleeve holding the prism.

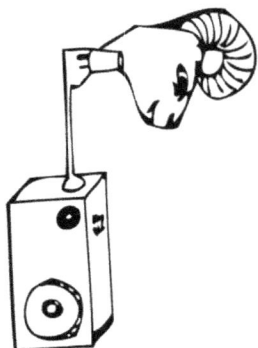

Remember - The "**RAM** it (**FLAT**)" rule of tonometry with high astigmatism. (See Illustration)

Place the **R**ed line of tonometer **A**t the **M**inus Axis
(i.e. on the axis of **FLAT**test corneal K).

AND

Those of you who use Harry Lorayne's alphanumeric encoding system will realize that the word "**RAM**" is equivalent to the number **43**. When the red line is placed on the minus axis it will split the ellipse at 43 degrees to the major axis.

DRIVING VISION REQUIREMENTS: SOME USEFUL APPROXIMATIONS

In most states (although this varies from state to state), the visual requirements for an unrestricted, non-commercial driving license for not-for-hire vehicles is granted if:

1) The patient has 20/40 or better visual acuity in the better eye, with or without corrective lenses.

AND

2) An uninterrupted visual field of 140 degrees horizontal diameter in one or both eyes.

40/40 guideline for driving vision

20/**40** vision AND
1**40** degrees of horizontal field (isopter unspecified)

*Note: All visual acuity and visual field requirements vary from state to state in the U.S. These are ONLY approximations.

Information source: 1. International Council of Ophthalmology, Vision Requirements for Driving Safety, Visual Standards Report. World Ophthalmology Congress, Sau Paulo, Brazil, February 2006. 2. Policy Statement, Canadian Ophthalmological Society Recommendations, Vision Standards for Driving in Canada, August 2000.

POWER CORRECTION IN RIGID CONTACT LENS FITTING

In RGP lens fitting, the power of the RGP must be adjusted depending on whether the lens is fit steeper or flatter than the measured K.

Remember – **SAMFAP** for RGP power correction.

SAM - If a lens is fit **S**teeper than K, **A**dd **M**inus power.
FAP - If a lens is fit **F**latter than K, **A**dd **P**lus power

Information source: American Academy of Ophthalmology Basic and Clinical Science Course: Clinical Optics. American Academy of Ophthalmology, 2008, p. 194.

| **TELESCOPES: ASTRONOMICAL VERSUS GALILEAN** | Astronomical telescopes consist of two plus lenses, separated by the sum of the focal lengths of the two lenses, and yield an inverted image. Galilean telescopes consist of a plus objective lens and a minus eyepiece, separated by the difference of the focal lengths of the two lenses, and yield an upright image. |

Remember – the properties of **As+**ronomical telescopes.

The letters "**A**" and "**t**" in the word **As+**ronomical tell you a lot about **As+**ronomical telescopes. First, the letter "**t**" looks a lot like a "**+**" sign, reminding you that **As+**ronomical telescopes have two "**+**" lenses separated by the "**+**" (or sum) of the focal lengths of the two lenses. Second, the letter "**A**" reminds you that they yield **A**ntiverted (inverted) images.

AND

Galilean telescopes yield "**G**ood" (upright) images.

| **OFFICE TESTS FOR UNILATERAL FUNCTIONAL VISION LOSS** | Some patients claim severe vision loss in the absence of any clinical signs. They may be malingering and have functional vision loss rather than an organic cause of their symptoms. Patients claiming unilateral visual loss, who retain visual acuity better than 20/400 may be especially difficult to diagnose and often leave the ophthalmologist flustered. Since these patients may also tend to be litigious, it is important to perform a careful exam. Some important clinical tests to include on your chart are listed below. |

Think - Malingerers are **PUFFERS**.

Because, malingerers and hysterics are full of "hot air," I like to think of them as **PUFFERS** when I perform a complete exam.

PUpil (relative afferent pupillary defect), **P**risms (fixation, diplopia)
Fog (plus lenses or plus-minus cylinder combinations off axis)
Filt ERS (duochrome, Polaroid vectograph)
Refraction (near vs. distance VA, walk-up, 20/10 line start)
Stereoacuity

| **EFFECT OF REFRACTION ON MEASURED OCULAR DEVIATION** | The prismatic effect of minus lenses will be an increase the measured strabismic deviation while plus lenses will decrease the measured deviation. |

Remember - **M**inus **M**easures **M**ore for the effect of minus lenses on measured strabismic deviation.

Mnemonic by permission of Elsevier. This article was published in Prisms in Optics and Refraction, Ophthalmology Board Review, Brent J. MacInnis, p. 37, copyright C.V. Mosby, 1994.

| **POSITIONING OF GLASS PRISMS** | Glass prisms are held in the *Prentice* position, with the posterior surface of the prism perpendicular to the line of sight. |

Think of a **GLASS-maker's apPRENTICE** to remember the *PRENTICE* position is used for **GLASS** prisms.

VISUAL ACUITY & STEREOPSIS

There is a definite correlation between binocular visual acuity and stereopsis. Testing of stereopsis can sometimes aid in the detection of non-organic vision loss.

mnemonic Adding the numerator and denominator of the Snellen acuity approximates the average stereopsis in seconds of image disparity. (See Table)

Visual Acuity	Approximation of Stereopsis	Average Stereopsis from Levy's Table
20/20	40	40
20/25	45	43
20/30	50	52
20/40	60	61
20/50	70	78
20/70	90	94
20/100	120	124
20/200	220	160

Information source: This table was reproduced with the permission of Elsevier.
This article was published in Stereoscopic Perception and Snellen Visual Acuity. Am J Ophthalmol, copyright Elsevier, vol 78: pp. 722-724, 1974.

CONTACT LENS BASE CURVE

A contact lens with a smaller radius of curvature (i.e. a smaller base curve) is a steeper lens.

mnemonic The **S**maller the base curve the **S**teeper the lens.
The **F**atter (bigger) the base curve the **F**latter the lens.

CORRECTING FOR ROTATION OF TORIC CONTACT LENSES

If the toric lens rotates to the left, the amount of rotation is added to the over-refracted axis. If the lens rotates to the right, the amount is then subtracted from the over-refracted axis.

mnemonic Remember the **LARS** rule of for rotation of toric contact lenses.

Left **A**dd & **R**ight **S**ubtract

Left **A**dd - If the toric lens rotates to the **L**eft, **A**dd the amount of rotation to the over-refracted axis.
Right **S**ubtract – If the toric lens rotates to the **R**ight, **S**ubtract the amount of rotation to the over-refracted axis.

("**RESLA**" is another version of this mnemonic)

An excellent graphic demonstration of toric contact lens fitting, utilizing the **LARS** mnemonic can be found in: Weinstock F, Contact Lens Fitting A Clinical Text Atlas. Gower Medical Publishing, 1989, p 49.

Mnemonic origin unknown.

| **DISPOSABLE CONTACT LENSES: SUREVUE VERSUS ACUVUE** | Surevue and Acuvue are registered trademarks of Johnson & Johnson. The Surevue lenses are daily wear lenses, and are thicker and easier to handle than the Acuvues. |

Remember - **S**urevues are **S**tiffer.

| **THE VISIBLE LIGHT SPECTRUM** | 1) The colors of the visible light spectrum, from longest wavelength to shortest are Red, Orange, Yellow, Green, Blue, Indigo, and Violet. |

ROY-G-BIV

Red, **O**range, **Y**ellow, **G**reen, **B**lue, **I**ndigo, and **V**iolet.

Mnemonic origin unknown.

2) Red has the longest wavelength in the visible spectrum.

Think of **Big Red** chewing gum to remember the **Big**gest wavelength of the visible spectrum is **Red**.

| **PHENOMENA WHEN LIGHT STRIKES A SURFACE** | When light interacts with a surface, there are several phenomena that may occur, including scatter, transmission, absorption, and reflection. |

Think of **STAR**light hitting a surface.

Scatter
Transmission
Absorption
Reflection

Mnemonic by permission of and modified from John A. Parker, MD, FRCS(C).

| **DIVERGING (CONVEX) MIRRORS: IMAGE CHARACTERISTICS** | When an incident beam of zero (or low vergence) hits diverging mirrors or diverging lenses the image is <u>smaller</u> than the object, <u>upright</u>, and <u>virtual</u> regardless of the distance of the object from the mirror or lens. (Note: This assumes is that the incident beam vergence is zero or low. Therefore, a strong convergent incident beam may form a real image if the mirror power is low.) To remember that the image is <u>smaller</u>, <u>upright</u>, and <u>virtual</u>: |

Think – **DI-V-ER**ging for diverging mirrors and lenses.

DIminished image (<u>smaller</u>)
Virtual image (<u>virtual</u>)
ERect image (<u>upright</u>)

NOTES & YOUR OWN MNEMONICS:

NOTES & YOUR OWN MNEMONICS:

III. ANTERIOR SEGMENT

TEAR FILM COMPONENTS

Classic dictum is that the outer layer of the tear film is lipid, the middle layer is aqueous, and the inner layer is mucin. To remember the layers of the tear film, remember:

mnemonic

The **O**uter is **O**il,
The **IN**ner is muc**IN**.

Thus – the **O**uter layer of the tear film is composed of **O**il
While the **IN**ner layer of the tear film is composed of muc**IN**.

DRY EYES

We all get bored with dry eye patients. However several diagnoses should be considered before discharging the patient with lubricating drops.

mnemonic

The patient's dry, *burning eyes* should set off **FIRE ALARMS**! (See Illustration)

Foreign body
Infiltrations
    (e.g. neoplasm/sarcoid of the greater superficial petrosal nerve, sphenopalatine ganglion, or lacrimal branch/gland)
Infections
    (e.g. herpes simplex & zoster, trachoma, TB, or leprosy)
Rosacea and other skin disorders such as ichthyosis
Exposure
    (secondary to lagophthalmos, thyroid, ectropion, or the 3 **P**'s of exposure, which are **P**roptosis, **P**alsy of CN 7, **P**SP,)
Avitaminosis **A**
Lash masquerade
    (e.g. entropion or trichiasis)
Aplasia of lacrimal gland
Autonomic dysfunction (especially in kids)
Rheumatoid arthritis
Medications
    (e.g. anticholinergics or birth control pills)
Mucous membrane diseases
    (e.g. OCP, chemicals, or erythema multiforme)
Sjogren's **S**yndrome
Systemic disorders
    (e.g. **S**arcoid or collagen vascular disease other than RA)
Spasm
    (e.g. blepharo**S**pasm)

CORNEAL DENDRITES

There is a wide differential for corneal dendrites that includes more than the obvious Herpes keratitis.

Think that the cornea **HATCHES Bar**s when you see corneal dendrites. (See Illustration)

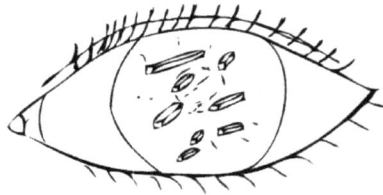

In a cornea with dendrites, imagine that the cornea **HATCHES Bar**s because corneal dendrites have a barbell shape.

Herpes simplex & zoster
Acanthamoeba
Tyrosinemia Type Two (Richner Hanhart syndrome)
Contact lens wear
Healing epithelial defect
Ep**S**tein-**Bar**r virus

HERBERT'S PITS

Herbert's pits are found in trachoma.

Think of the racing Love Bug "**Herbie**" (Herbert) making a **Pit** stop at the race **track** (trachoma).

PAPILLARY CONJUNCTIVITIS

When evaluating conjunctivitis, the identification of certain clinical features can aid in creating the differential diagnosis. Papilla (central dilated, telangiectatic conjuntival vessels surrounded by peripheral edema) are commonly found in allergic or bacterial conjunctivitis. The differential is short.

Think – **PAB**illary conjunctivitis whenever you see papilla.

That's **P**apillary conjunctivitis = **A**llergic or **B**acterial conjunctivitis.

FOLLICULAR CONJUNCTIVITIS

The differential for conjunctival follicles (central lymphoid nodules surrounded by peripheral accessory vascularization) is much longer than that of papilla. Follicles are commonly found in adenovirus, molluscum contagiosum, herpes simplex, drug/drop reaction, and chlamydial conjunctivitis.

If you're not good like the **AMiSH,** you may suffer from the **V.D. Clam Follies**.

Think of **VD** (venereal disease) such as **Clam** (Chlamydia) being the result of **folly**.

Adenovirus
Molluscum contagiosum
i
Simplex (Herpes)

Viral (including Adenovirus, Molluscum, Herpes simplex**)**
Drops (**D**rugs)

Clam (Chlamydia)
Follies (follicular conjunctivitis)

| **CLINICAL FINDINGS IN KERATOCONUS** | There are many exam findings that can help you with the diagnosis of keratoconus. |

In keratoconus, the cornea is in "**SHAB V FoRM**"

"**SHAB V FoRM**" sounds like "Shabby Form."
"**V**" indicates the conical corneal protrusion of keratoconus.

Scissoring on retinoscopy	**F**leischer ring
Hydrops	o
Astigmatism	**R**izzuti's sign
Breaks (in Descemet's and Bowman's)	**M**unson's sign
Vogt's striae	

KERATOCONUS: RISK FACTORS

Several conditions are thought to predispose patients to the development of keratoconus.

An **Atopic EOM RubS Down** causing keratoconus.

Imagine that an itchy, **Atopic EOM** (Extra**O**cular **M**uscle) **RubS Down** on the cornea to causing it to warp and develop keratoconus.

Atopic disease (asthma, eczema, atopic keratoconjunctivitis, & hay fever)
Ehlers-Danlos
Osteogenesis imperfecta
Marfans syndrome
Rubbing of eye
Sleep apnea
Down's syndrome

KERATOCONUS: TREATMENT OPTIONS WHEN GLASSES DON'T HELP

Vision rehabilitation in keratoconus initially consists of spectacle correction of the patient's refractive error. However, as the disease progresses, adequate vision may not be obtained with spectacles alone, and other treatment modalities must be considered.

Think of the astigmatic **SHIFT** to remember treatment options for keratoconus.

Soft contacts
Hard contacts (RGP)
Inserts (corneal inserts such as Intacs)
Flavin (corneal collagen crosslinking with riboflavin)
Transplant

VOGT'S STRIAE: DEFINITION

Vogt's striae are vertical stress lines commonly seen in keratoconus.

Think - **V**ogt's **STR**iae are **V**ertical **STR**ess lines in the **STR**oma.

CORNEAL WHORLS

There is rather broad differential for corneal whorls, and this is a popular question on exams and on rounds. To remember the differential:

Think – **"PICA FaST"** for corneal **whorls**.

After eating dirt (a.k.a. **PICA**) you will be sick (see **whorls**) and be sick **FaST**. Remember that **PICA** ("pica") is medical disorder characterized by an appetite for largely non-nutritive substances such as dirt.

Phenothiazines	**Fa**bry's disease
Indomethacin	**S**uramin†
Chloroquine & hydroxy**C**hloroquine	**T**amoxifen
Amiodarone	

†Suramin is an anti-parasitic drug used in oncology and HIV patients.

Mnemonic origin unknown.

CORNEAL IRON LINES

The different corneal epithelial iron lines are a favorite trivia question. Hudson-Stahli lines are commonly seen in older patients as a horizontal line just inferior to the center of the interpalpebral fissure. Stocker lines are seen at the leading edge of a pterygium. Fleischer rings are seen at the base of the cone in keratoconus. Ferry lines are seen at the corneal edge of a filtering bleb. There are many other eponymous iron lines. Remember your corneal iron lines with this little rhyme and some visual imagery.

**Filter Ferry, Fleischer* Kerry,
Stocker Ptery†, H.S. Geri.**

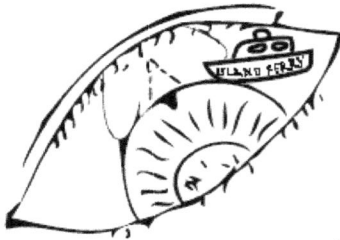

That is to say, **Filter**ing blebs form **Ferry** lines. (See Illustration)
Fleischer rings* occur in **Kerry**-toconus (keratoconus).
Stocker lines occur in **Ptery**†-gium.
and **H.S.** (**H**udson-**S**tahli) lines occur in **Geri**-atric (older) patients.

*For the Fleischer Ring

Think of carrying round tub of Fleishmann's margarine.

"carrying" = keratoconus
"Fleischmann" = Fleischer
"round tub" = ring

†For the Stocker Line

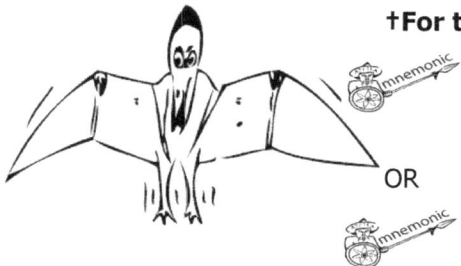

"Ptery"gium sounds like **"ptery"**dactyl and I imagine a pterodactyl **stalk**ing its victims. (See Illustration)

OR

Alternatively you can think of **"terrible stocks."**

BAND KERATOPATHY

Band keratopathy is characterized by calcium deposition in the cornea.

KAL-SI-UM DeP-osits cause band keratopathy.

KALSIUM (hypercalcemia)
Kidney failure (causing hyperphosphatemia)
ALkali (milk alkali syndrome)*, **A**rthritis (juvenile rheumatoid arthritis)
Sarcoidosis*, **S**ilicone oil, **S**kin problems (e.g. ichthyosis)
Inflammation of the chronic type (e.g. juvenile rheumatoid arthritis & phthisis)
Uric acid elevation (gout)
Mercury vapor, **M**iotics (add **M**ilk alkali syndrome here if you forget what "**AL**" stands for above)

D vitamin toxicity*
Parathyroidism (hyper)*

*Causes of hypercalcemia include: milk alkali syndrome, sarcoidosis, D vitamin toxicity, and hyperparathyroidism.

Note: Juvenile arthritis is repeated twice, so it won't be forgotten.

CORNEAL CRYSTALS: DIFFERENTIAL DIAGNOSIS

The associations of corneal crystalline deposits are another popular board and quiz question with a wide differential.

Think - **CRISTAL BUMPS** cause corneal crystals.

Cystinosis, **C**iprofloxacin
Roids (steroid drops)
Infection with *Strep viridans*
Schnyder's crystalline dystrophy
Tyrosinemia
Argyrosis
Lipid, **L**CAT deficiency, **L**owe's

Bietti's, **B**and keratopathy
Uric acid (gout)
Myeloma*
Paraproteinemia*, **P**orphyria cutanea tarda
Sap from the Diffenbachia plant

*Myeloma and paraproteinemia are almost identical, but repeated since they are important not to miss.

Information source: American Academy of Ophthalmology Basic and Clinical Science Course: External Diseases and Cornea. American Academy of Ophthalmology, 2008, pp. 370-378.

ENLARGED CORNEAL NERVES: DIFFERENTIAL DIAGNOSIS

Enlarged corneal nerves can be due to inherited, infectious, and local or systemic disease. To remember the many causes of enlarged corneal nerves:

Think of **MEN₂LARGED** corneal nerves which sounds like "Enlarged" corneal nerves.

MEN2 Multiple Endocrine Neoplasia II B (II A, and III also)
Neurofibromatosis (the second "N" in **MEN₂**)
Leprosy
Acanthamoeba
Refsum's disease (phytanic acid storage disease)
Glaucoma (congenital glaucoma (makes corneal nerves more visible))
Edema (corneal edema also makes corneal nerves more visible)
Dysautonomia (familial dysautonomia a.k.a. the Riley-Day syndrome)

Information source: American Academy of Ophthalmology Basic and Clinical Science Course: External Diseases and Cornea. American Academy of Ophthalmology, 2008, pp. 357-358.

PEMPHIGUS VERSUS PEMPHIGOID

Pemphigus vulgaris is a skin disease that rarely affects the eyes and does not cause conjunctival scarring. Pemphigoid can cause a chronic cicatrizing conjunctivitis. Therefore, as ophthalmologists. . .

Pemphig**OID** we want to AV**OID**, and
Pemphig**US** is better for **US**.

CHARACTERISTICS OF OCULAR CICATRICIAL PEMPHIGOID (OCP)

OCP is a type II (cytotoxic) hypersensitivity reaction and the bullae are subepithelial. For this reason scarring occurs. Remember this little rhyme for OCP:

OCP is Sub-Epi.

Say this mnemonic while pointing two fingers at each letter to remind you that OCP is a type II hypersensitivity reaction – "**OCP is Sub-Epi.**" (See Illustration)

You can also imagine **sub**-marines under each letter of OCP reminding you that OCP has **sub**-epithelial bullae.

CORNEAL PRECIPITATES DUE TO CIPROFLOXACIN

Frequent administration of ciprofloxacin in corneal ulcer patients may lead to white crystalline precipitates in approximately 17% of patients. Ofloxacin (as well as newer fourth-generation quinolones) do not cause precipitates, presumably because of their better solubility.

ci**PR**ofloxacin **PR**ecipitates.

WILSON'S DISEASE: BLOOD TESTING

Serum ceruloplasmin is usually markedly reduced (a.k.a. "**LOW**") in Wilson's disease.

Remember - ceru**LOW**plasmin is **LOW** in Wilson's disease.

CLASSIC STROMAL CORNEAL DYSTROPHIES

The corneal dystrophies are heritable corneal diseases associated with the deposition of abnormal material within the cornea. Each of the "classic" corneal dystrophies is characterized by the material deposited in the cornea and the unique stain for that material.

Marilyn Monroe Always Got Her Man in L.A. County.

Dystrophy	Deposit Material	Stain
Macular	**M**ucopolysaccharide	**A**lcian blue
Granular	**H**yaline	**M**asson trichrome
Lattice	**A**myloid	**C**ongo red

Mnemonic origin unknown. It can be found in J. A. McWhae's "Acronyms and Mnemonics in Ophthalmology."

MACULAR DYSTROPHY: PROGNOSIS

In distinction to the other corneal dystrophies, macular dystrophy is autosomal recessive and leads to a decrease in vision at an earlier age. Macular dystrophy is associated with disorders of keratan sulfate metabolism.

Macular dystrophy **MAR-KS** the cornea the most.

Macular dystrophy
Autosomal **R**ecessive
Keratan **S**ulfate

Note: Type II macular dystrophy is also associated with a disorder of dermatan sulfate-proteoglycans.

REIS-BÜCKLERS DYSTROPHY

Reis-Bücklers dystrophy mainly affects Bowman's layer.

Remember – Reis-**B**ücklers affects **B**owman's layer.

Mnemonic by permission of and modified from Michael Myles, MD, FRCS(C).

ORDER OF CULTURE PLATE INOCULATION WITH CORNEAL SCRAPING

When culturing infectious corneal lesions, the blood agar plate should be inoculated first, since the corneal specimen may be limited and blood agar supports growth of the widest spectrum of corneal pathogens.

Be aggressive & "**draw first blood**" when performing cultures for bacterial corneal ulcers.

In other words, **draw** the "C" streaks or zigzags **first** on the **blood** agar plates when culturing bacterial corneal ulcers.

PURULENT CONJUNCTIVITIS: H. FLU BIOTYPE III VS. PNEUMOCOCCUS	The acute purulent conjunctivitis of *Haemophilus influenzae* biotype III (formerly called *Haemophilus aegyptius*) resembles that of *S. pneumoniae*. However, pneumococcal infections may have a conjunctival membrane, while *Haemophilus influenzae biotype III* usually does not cause a conjunctival membrane. Remember the distinction with this little rhyme:

H. flu *three*
Is membrane *free*.

Information source: American Academy of Ophthalmology Basic and Clinical Science Course: External Diseases and Cornea. American Academy of Ophthalmology, 2008, p. 170

ORGANISMS THAT CAN PENETRATE AN INTACT CORNEAL EPITHELIUM	Certain bacteria, most commonly gonococcus (but also *Haemophilus influenzae* biotype III, *Corynebacterium diphtheria*, and *Listeria monocytogenes*) can penetrate an intact corneal epithelium. This phenomenon is known as epithelial parasitism. Remember the organisms capable of penetrating an intact corneal epithelium with this little rhyme:

3 Ham, Corn dip* isn't **Nice** on my **List,**
It's a recipe for spontaneous dehisce.

Corn dip reminds you of two things
      1. The **Corneal dip**s (excavations) of the infectious ulcer
      2. The organism, *Corynebacterium diphtheria*

3 Ham	*Haemophilus influenzae* biotype III
Corn dip	*Corynebacterium diphtheriae*
Nice	*Neisseria gonorrhea* (and *meningitidis*)
List	*Listeria monocytogenes*

CANDIDA KERATITIS	The fungus, *Candida* is an example of a yeast that causes fungal keratitis (as opposed to the other groups of fungi – the filamentous and dimorphic fungi).

Candida causes **Cand-Y** keratitis that's best treated with **C-Ampho**r.

"**Cand-Y** keratitis" reminds you that **Cand-**ida is a **Y**east.

"best treated with **C-Ampho**r" reminds you that Candida is best treated with topical **Ampho**-tericin.

Note: Camphor itself, also happens to have anti-infective properties.

FILAMENTOUS FUNGAL ULCERS	Fungal ulcers following trauma are usually due to filamentous fungi which are more common in the warmer climates of the southern U.S. In contrast, yeast infections are more common in the northern U.S. and often seen in eyes with pre-existing disease.

Think of **crushing** a **hot** light bulb **filament** into someone's cornea causing a traumatic fungal ulcer.

Crushing reminds you of trauma.
The **filament** reminds you that filamentous fungal ulcers typically occur after this trauma.

"**Hot**" in **hot** light bulb **filament**, reminds you that filamentous fungal ulcers are more common in the "**hot**" weather states in the southern U.S.

ACANTHAMOEBA CORNEAL ULCER

The diagnosis and treatment of Acanthamoeba corneal ulcers is unique among the many other causes of microbial keratitis. To help you remember this unique condition:

mnemonic   Remember – The Seven **C**'s of '**C**anthamoeba

Diagnosis
    Contacts (often contact lens related)
    Calcofluor white (one of the stains for acanthamoeba)
    Coli overlay of non-nutrient agar (E. coli overlay is the preferred culture medium for Acanthamoeba)
Treatment
    Clotrimoxazole (topical)
    Conazoles (oral ketoconazole or fluconazole)
    Chlorhexidine
Prevention
    Contact lens cleaner

ACANTHAMOEBA: TOPICAL MEDICATIONS

Acanthamoeba is an exceptionally difficult to treat cause of microbial keratitis. It requires the use of unique medications and topical preparations, often in combination and used for months, in an attempt to eradicate the organism.

mnemonic   Remember - **POLLY**'s **BIG CLOT**hes **PI-NS** to treat Acanthamoeba.

Pretend that you will squeeze and eradicate the Acanthamoeba cysts with **POLLY**'s **BIG CLOT**hes **PI-NS**.

POLLY's **BIG**	**POLY**hexamethylene **BIG**uanide (a pool cleaner)
CLOThes	**CLOT**rimoxazole
PI	**P**ropamidine **I**sethionate (trade name Brolene)
NS	**N**eo**S**porin (which contains **N**eomycin)

NEOSPORIN DROPS VERSUS NEOSPORIN OINTMENT

The drug trade named Neosporin contains Polymyxin B and Neomycin. In addition, the drop form contains Gramicidin, while the ointment form contains Bacitracin.

mnemonic   Neosporin **G**tts contain **G**ramicidin.

The abbreviation for drops is "**G**tts" and the Neosporin **G**tts contain the **G**ramicidin (while the ointment contains Bacitracin).

CHARACTERISTICS OF SESSILE CONJUNCTIVAL PAPILLOMAS

As opposed to pedunculated conjunctival papillomas, sessile conjunctival papillomas are characterized by:

mnemonic

The Six "**S**'s" of **SeSS**ile papillomas.

Single
Sinister (malignant potential)
Senior patients have them more often (i.e. more adults vs. children)
Scleral **S**ulcus location (i.e. more common at the limbus)
Sixteen (sessile papillomas are associated with Human Papilloma Virus type 16)

RADIAL KERATOTOMY

Although radial keratotomy (RK) is not routinely performed as it once was, the procedure is still of historical interest.

mnemonic

Rule of **4**'s for radial keratotomy.

4 initial cuts for myopia **4** D or less of myopia
4 mm central clear zone should be left (3-5mm)
40% rate of undercorrection and re-operation is tolerated in most algorithms

Information source: American Academy of Ophthalmology Basic and Clinical Science Course: External Disease and Cornea. American Academy of Ophthalmology, 1994, p. 335.

PHACOEMULSIFICATION VACUUM SYSTEMS

Current phacoemulsification vacuum systems include the venturi pump and the peristaltic pump. Diaphragmatic pumps are no longer used in modern phacoemulsification machines, and are mainly of historical interest. The quick rise time of venturi system is desired by many experienced surgeons. However, beginning cataract surgeons may prefer the slower rise of a peristaltic vacuum system, finding the venturi vacuum moves too fast for their skill level.

mnemonic

A **VENTUR**i systems may lead to an unwanted ad**VENTUR** for the beginning cataract surgeon.

mnemonic

DIEaphragmatic systems have **DIE**d and are mainly of historic interest.

EFFECTIVE INTRAOCULAR LENS POWER

When contending with vitreous during cataract surgery (i.e. a ruptured posterior capsule), you want to avoid becoming flustered and choosing the wrong lens power for the sulcus or anterior chamber placement.

mnemonic

Remember - For a lens in the **SU**lc**US**, **SU**btract pl**US**.

OR

mnemonic

For **DECREASE**d depth of lens insertion (i.e. in the sulcus or AC) **DECREASE** the IOL power.

| **PROPERTIES OF VISCOELASTICS** | Viscoelastics used in modern cataract surgery vary by their viscosity, molecular weight, and cohesiveness (ease with which they are aspirated from the eye). |

1) <u>Sodium hyaluronate</u> is of high viscosity, high molecular weight, and is easily aspirated from the anterior chamber.

Sodium **HIGH**aluronate is of **HIGH** viscosity, **HIGH** molecular weight, and **HIGH**tails it out of the eye.

2) <u>Chondroitin sulfate</u> is of medium viscosity and medium molecular weight.

Think **CSM** for chondroitin sulfate.

Chondroitin **S**ulfate is of **M**edium viscosity and **M**edium molecular weight.

"**CSM**" is a commonly used abbreviation in pediatric ophthalmology.

3) <u>Hydroxypropyl methylcellulose</u> is of low viscosity and low molecular weight.

Think methylcellu**LOW**se is of **LOW** viscosity and **LOW** molecular weight.

| **DISPERSIVE VISCOELASTICS** | Dispersive viscoelastics include Viscoat (Alcon) and Vitrax (AMO). These dispersive viscoelastics can be injected over a posterior capsule tear to prevent the vitreous from moving anteriorly. |

VD's are easily dispersed.

Just as **VD**'s (venereal diseases) are easiy dispersed, the viscoelastics **V**iscoat and **V**itrax are easily **D**ispersed within the eye.

V = **V**iscoat and **V**itrax
D = **D**ispersive viscoelastics

AND

Make a "**V**" with your arms (i.e. "**disperse**" them) to remember that **dispersive** viscoelastics start with a "**V**."

| **PHACOEMULSIFICATION HANDPIECES** | Phacoemulsification handpieces may be cooled using different mechanisms designed into the handpiece. The MicroFlow® handpiece uses fluid running through furrows along the length of the shaft for cooling. The MicroSeal® handpiece uses a low friction dual sleeve technology to prevent heat buildup. |

To cool the handpiece:
Micro**F**low uses **F**urrows,
Micro**S**eal uses a dual **S**leeve .

Note: both MicroFlow® and MicroSeal® are trademarks of Storz Ophthalmics.

PHACOEMULSIFICATION: THE RUPTURED POSTERIOR CAPSULE

When the posterior capsule is ruptured during phacoemulsification, certain maneuvers may be employed in an attempt to "save" the surgery and produce a satisfactory result. When the posterior capsule is ruptured:

Remember - The "**PAL SAVE.**"

These management options are like a "**PAL**" that might help "**SAVE**" the day when the posterior capsule is ruptured.

PAL = **P**osterior **A**ssisted **L**evitation (as first described by Kelman, instruments are passed through pars plana to "levitate" the nuclear fragments into the anterior chamber)

Sheets glide (to place over the rent)
Simcoe needle (for manual I & A)
Stain the vitreous with triamcinolone (trade name Kenalog) for vitrectomy
Attenuate the infusion (lower the bottle)
Viscoelastic* (to hold back the vitreous)
Vitrectomy
Extracapsular conversion (possibly, if the rupture is large or is enlarging)

*Note: Many surgeons advocate for the injection of viscoelastic immediately (before stopping infusion and before removing the phacoemulsification handpiece from the eye) when a rupture in the posterior capsule is noted. This is in an effort to prevent extension of the posterior capsular tear by sudden changes in pressure.

BIGH3-RELATED CORNEAL DYSTROPHIES

The linkage of Reis-Bücklers and other corneal dystrophies (Granular, Avellino, and Lattice) to 5q31 has been reported. TGFBI (transforming growth factor - beta-induced), initially called BIGH3, encodes for a 683 amino acid protein product, with several nomenclatures including "keratoepithelin" and the now the preferred name "TGFBI associated protein" (abbrviated TGFBIp).

Remember – "**BIG** is **LARG**e" for the BIGH3-related corneal dystrophies.

BIG = BIGH3
Lattice
Avellino
Reis-Bücklers
Granular
e

Mnemonic origin unknown. Special thanks to Anthony Johnson MD for mentioning it.

MECHANISM OF PAIN IN POST-HERPETIC NEURALGIA

Post-herpetic neuralgia is a chronic and/or intermittent pain syndrome that follows an acute eruption of Herpes simplex or Herpes zoster.

mnemonic

Remember- The **ABC's** of post-herpetic neuralgia.
It's caused by **A**ngry, **B**ack-firing, **C**-nociceptors.

Mnemonic by permission of Oxford University press from M Cline et al, Brain, 1989, Vol 112, copyright 1989 Guarantors of Brain, pp. 621-647.

NOTES & YOUR OWN MNEMONICS:

NOTES & YOUR OWN MNEMONICS:

IV. UVEITIS

TYPES OF HYPERSENSITIVITY: THE GELL & COOMBS CLASSIFICATION

Although the adaptive immune-triggered inflammatory responses of the Gell & Coombs system were conceived before T-lymphocytes were discovered, and the "hypersensitivity" may actually be protective or non-inflammatory, this classification system is still in use.

Hypersensitivity Type	Mediator
Type 1	Anaphylactic
Type 2	Cytotoxic antibody mediated
Type 3	Immune Complex mediated
Type 4	Cell mediated
Type 5*	Stimulatory

Remember - **AN-CY IMC-CE-ST** for hypersensitivity reactions.

We all become a little **AN-CY** (i.e. antsy or "hypersensitive") when we hear about **IM-CE-ST** (sounds like "incest").

ANaphylactic –	type 1
CYtotoxic –	type 2
IMC = **IM**mune **C**omplex –	type 3
CEll mediated –	type 4
STimulatory –	type 5*

*Note: Type 5 hypersensitivity is recognized by some sources.

Information source: American Academy of Ophthalmology Basic and Clinical Science Course: Intraocular Inflammation and Uveitis. American Academy of Ophthalmology, 2008, p. 54.

IRIS NODULES IN UVEITIS

There are two types of iris nodules found in uveitis, Busacca's nodules and Koeppe's nodules, and these are a commonly quizzed topic in ophthalmology. Busacca's nodules are found on the iris stroma and are usually associated with granulomatous iritis, while Koeppe's nodules are located at the pupil margin (and not specific for granulomatous iritis).

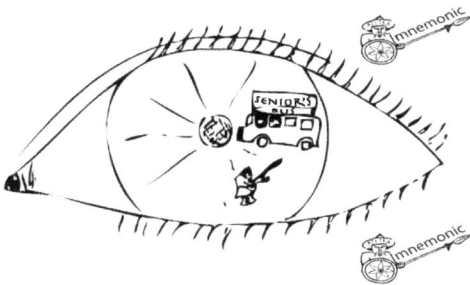

For **Bus**accas' nodules think of a **Bus**load of **Gra**nnies!
(See Illustration)
Imagine that a **Bus** carrying a **Bus**load of **Gra**nnies must sit firmly on the iris stroma (if the **Bus**load of **Gra**nnies was on the pupil margin it would fall over). This **Bus**load of **Gra**nnies also reminds you that **Bus**acca's nodules are usually associated with **Gran**ulomatous iritis.

For **K**oeppe's nodules remember that **K**oeppe's are near the "**K**ollar."

In other words, **K**oeppe's nodules are located at the iris "**K**ollar" (the "collar" or edge of the iris at the pupil margin).

Mnemonic by permission of and modified from Dave Chacko, MD, PhD.

VITAMIN REPLACEMENT FOR PATIENTS ON PYRIMETHAMINE

Pyrimethamine (trade name Daraprim) is one antibiotic that may be used in the treatment of ocular toxoplasmosis. This treatment requires vitamin replacement with oral folinic acid (NOT folic acid).

Foli**NIC** acid is the **NIC**e way to go with pyrimethamine treatment.

HLA B27 AND IRITIS

HLA B27 is usually associated with an acute, anterior, non-granulomatous iritis. There is a 25% chance that a patient with HLA type B27 will develop back or eye problems.

Remember - The words "HL**A B27**" tell much.

A   **A**cute, **A**nterior iritis
B   The association with **B**ack disease (i.e. ankylosing spondylitis)
27 It affects patients about **27** years old (usually **not grannies**, thus **non-granulomatous** uveitis) and there is approximately a **27%** risk of back or eye problems.

Information source: American Academy of Ophthalmology Basic and Clinical Science Course: Intraocular Inflammation and Uveitis. American Academy of Ophthalmology, 2008, p. 150.

THE "SIXES" OF SYPHILIS

Treponema pallidum is the spirochete that causes the infectious disease syphilis.

Remember the many **"Sixes" of Syphilis**.

Treponema pallidum has **6** to 14 spirals and is **6** to 20 micrometers long.

The primary lesion (a chancre) appears within 2-**6** weeks of exposure, persists for **6** weeks, and then heals spontaneously.

6 weeks after healing of the primary lesion, secondary manifestations appear. Initially these secondary lesions are **6** mm macules, which after **6** weeks become **6** mm papules.

Symptomatic neurosyphilis occurs in about **6**% of untreated syphilitics.

16% of untreated cases develop "benign" tertiary disease (gumma formation).

Aortitis occurs in up to **60**% of autopsy cases.

Information source: J. Lawton Smith, in Neuro-Ophthalmological Disorders, RJ Tusa & SA Newman (eds), Informa HealthCare, 1995, p. 546.

Mnemonic origin unknown. Special thanks to J. Lawton Smith, MD for mentioning it.

HLA A29 & BIRDSHOT RETINOCHOROIDOPATHY

Birdshot retinochoroidopathy is associated with HLA A29 in 80-98% of patients.

Picture a flamingo bird constructed from an **A**, **2** and **9**.
(See Illustration)
The **A** on its side is the flamingo's beak.
The **2** forms the neck and body.
The **9** is the leg.

Notice that much of the flamingo is missing since it's been **shot** – reminding you of the association between bird-**shot** retinochoroidopathy and HLA **A29**.

Information source: American Academy of Ophthalmology Basic and Clinical Science Course: Intraocular Inflammation and Uveitis. American Academy of Ophthalmology, 2008, p. 180.

MANIFESTATIONS OF REITER'S SYNDROME	Patient's with Reiter's syndrome present with conjunctivitis, urethritis, and polyarthritis. To remember the manifestations of Reiter's syndrome:

Think of a "**Writer**" (Reiter) and his coffee **CUP** — The **Writer's CUP**!

Writer = Reiter's Syndrome
Conjunctivitis
Urethritis
Polyarthritis |
| **CAUSES OF SECONDARY GLAUCOMA IN UVEITIS** | There is a wide differential diagnosis to consider for secondary glaucoma in the uveitis patient.

Think - **DeBRIS PAST**e**S** the trabecular meshwork in uveitic glaucoma.

Debris
Bombe of the iris
Rubeosis **I**ridis
Steroid-induced glaucoma

Peripheral **A**nterior **S**ynechiae
Trabeculitis (inflammation of the trabcular meshwork)
e
Sclerosis of trabecular meshwork |
| **CONGENITAL INFECTIONS** | Many congenital infections passed from the mother to fetus may have ocular manifestations. These are often referred to as the "**ToRCHeS** complex" of congenital infections.

The **ToRCHeS** complex of congenital infections.

Toxoplasmosis
Rubella
CMV
Herpes simplex
Syphilis

Mnemonic origin unknown. |
| **TOXOPLASMOSIS** | Toxoplasmosis in neonates may manifest with retinochoroiditis, convulsions, intracranial calcifications, and cerebrospinal fluid abnormalities.

Remember - The **6 C's** of **C**NS toxoplasmosis.

Chorioretinitis
CNS **C**alcifications best seen on **CT**
Convulsions
CSF abnormalities

Mnemonic by permission of and modified from Mike Myles, MD, FRCS(C). |

NOTES & YOUR OWN MNEMONICS:

V. ORBIT & OCULOPLASTICS

THE OPTIC FORAMEN: LESSON "A" OF ORBITAL OSTEOLOGY

The optic foramen is located in the lesser wing of the sphenoid bone and contains the all important optic nerve (and its surrounding meninges), as well as the ophthalmic artery and the sympathetics.

As a neuro-ophthalmologist and orbital surgeon, the first and most important lesson in orbital osteology is the location of the optic foramen, since it houses the optic nerve. Therefore I designate the location and contents of the optic foramen as:

mnemonic → **LESSON "A" OF** orbital osteology.

LEsser wing of **S**phenoid
Sympathetics
Optic **N**erve (and its meninges)
Artery (ophthalmic artery)
Optic **F**oramen

NERVES OUTSIDE THE OCULOMOTOR FORAMEN

The portion of the orbital apex enclosed by the annulus of Zinn is the oculomotor foramen. The lacrimal and frontal divisions of CN V, and the trochlear nerve travel in the superior orbital fissure, but outside of the oculomotor foramen.

mnemonic → The **L**acrimal, **F**rontal, and **T**rochlear nerves are **LeFT** out of the oculomotor foramen.

CONTENTS OF SUPERIOR ORBITAL FISSURE

Many structures pass through the superior orbital fissure, and these can be difficult to remember. These structures include cranial nerves 3, 4, V1 and 6, as well as the sympathetics, the superior ophthalmic vein, and the anastomosis of recurrent lacrimal & middle meningeal arteries. To remember the contents of the superior orbital fissure, think of this little rhyme:

mnemonic → **Three, Four, S.O., V1, & Six,**
Plus **anastomosis** go through the **SO Fix**.

Three	CN 3
Four	CN 4
S.	Sympathetics
S.O.	Superior Ophthalmic vein
V1	CN V1
Six	CN 6
Plus **anastomosis**	Anastomosis of recurrent lacrimal & middle meningeal arteries
go through the **SO Fix**	Superior Orbital Fissure

Cranial nerves three, four, sympathetics, superior ophthalmic vein, cranial nerve V1, and cranial nerve six run through the superior orbital fissure as well as the anastomosis of recurrent lacrimal & middle meningeal arteries.

ORBITAL OSTEOLOGY

The bony orbit is a pyramidal structure with the orbital apex forming the peak of the pyramid. The four bony walls of the orbit, constructed of various cranial bones, are the medial and lateral walls, the roof and floor. To help organize your thinking, remember that the sphenoid bone (through its greater and lesser wings) is thought to be the "keystone" of the orbit and contributes to all of the orbital walls except the floor.

Medial Wall: The medial wall of the orbit composed of 4 bones: the Sphenoid, Maxillary, Ethmoid, & Lacrimal

Think – **SMEL** for the bones of the medial wall.
(See Illustration)
The medial wall is close to the nose, the organ of **SMEL** ("smell").
Bones of the medial wall – **S**phenoid, **M**axillary, **E**thmoid, & **L**acrimal

Mnemonic origin unknown. It can be found in J. A. McWhae's "Acronyms and Mnemonics in Ophthalmology."

Orbital Floor: The floor of the orbit is composed of 3 bones: the Maxillary, Palatine, & Zygomatic.

Think - **MaPZ** or floor **MoPZ** for the bones of the orbital floor.
(See Illustration)
Imagine that you spread **MaPZ** ("maps") out on orbital floor to read them, or imagine mopping the orbital floor with floor **MoPZ** (floor "mops").
Bones of the orbital floor – **Ma**xillary, **P**alatine, & **Z**ygomatic

AND

The palatine bone is the most posterior bone of the orbital floor.

Remember – The **P**alatine bone is most **P**osterior bone of the orbital floor.

Lateral Wall: The lateral wall of the orbit is composed of 2 bones: the Zygomatic & Greater wing of the sphenoid. Anatomically, it's easy to remember the Zygomatic is part of the lateral wall. If you need a little help:

Think – The **Late GreatZ** for the bones of the lateral wall.
(See Illustration)
Imagine seeing a marquee advertising the **Late GreatZ** (those great movies of yesteryear) flashing in the corner of your eye while studying this book.
Bones of the **Late**ral wall – **Great**er Wing of Sphenoid & **Z**ygomatic.

Orbital Roof: The roof of the orbit is composed of two bones: the Frontal & Lesser wing of the sphenoid.

Think – **RooFLess** for the bones of the orbital roof.
(See Illustration)
If you removed these two bones, the orbit would be "**RooFLess**!"
Bones of the orbital **RooF** – **F**rontal bone & **Less**er wing of Sphenoid.

OR

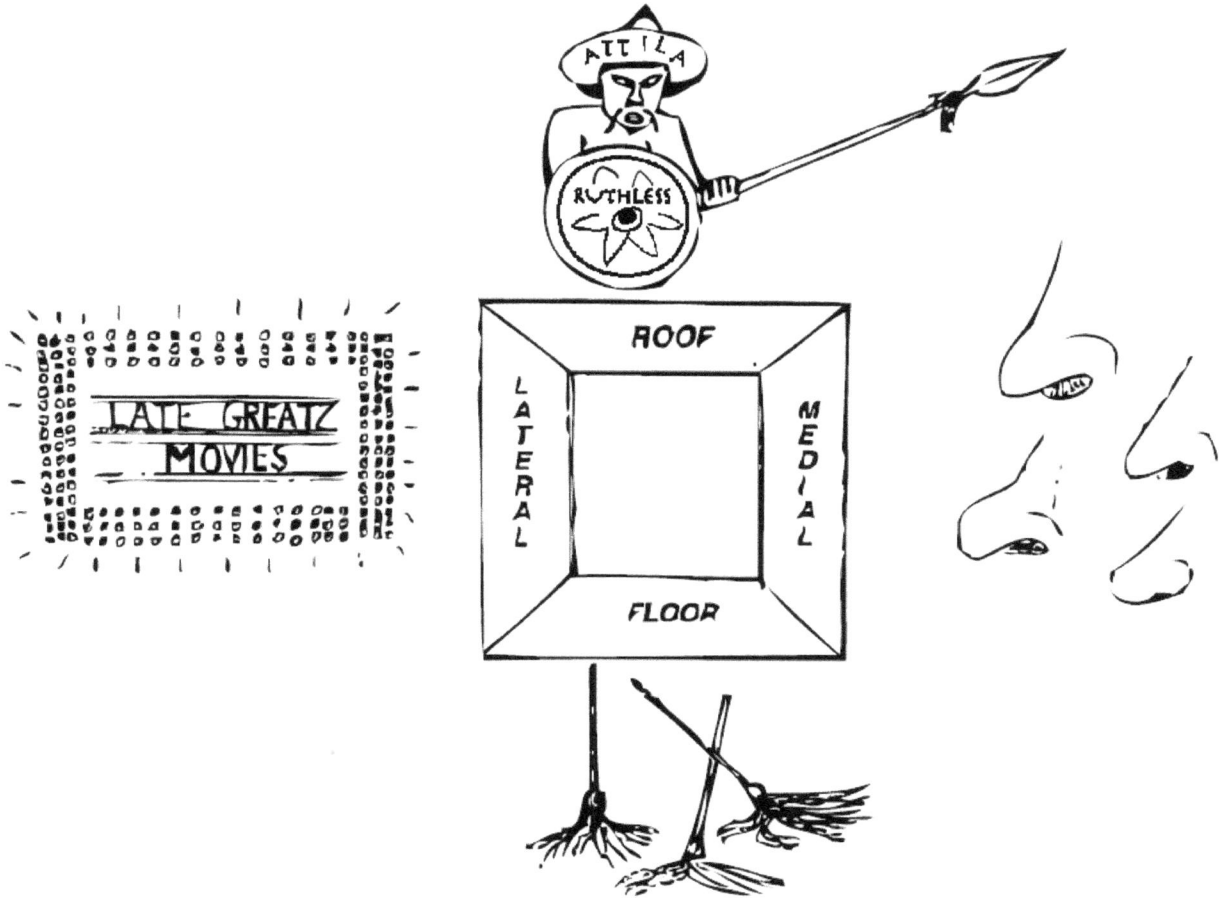

Alternatively think of Attila the Hun sitting atop the orbit.
Attila was a **RooFLess** ("ruthless") warrior.

**ABNORMAL SIZE OF THE
OPTIC FORAMEN**

In young children, an optic foramen greater than 6.5 mm in diameter is
considered abnormal. Additionally, asymmetry of the optic foramina greater
than 1 mm is also abnormal.

An optic foramen in children **6.6** mm (or greater) is **sick-sick**!

The number **6.6** ("six point six") sounds like "sick sick" – reminding
you that an optic foramen in children 6.6 mm or greater is abnormal.

Note: In adults an optic foramen that measures 7 mm in diameter is abnormal.

Information source: American Academy of Ophthalmology Basic and Clinical Science Course: Orbits,
Eyelids and Lacrimal System. American Academy of Ophthalmology, 1992, p. 19.

MEDIAL ORBITAL LANDMARKS: DISTANCES

When dissecting along the medial orbit, it is important to know the expected distances to the ethmoidal vessels and the optic foramen. The anterior and posterior ethmoidal arteries pass through the corresponding ethmoidal foramina.

mnemonic **The Rule of 24** for the structures of the medial orbital wall.

If you don't know the location of the anterior ethmoidal artery when dissecting along the medial wall, you'll spend **24** hours trying to stop the bleeding. The distance to each subsequent foramen is then halved.

Structure	Distance
From ant. lacrimal crest to ant. ethmoid foramen (containing the ant. ethmoidal artery)	**24** mm
From ant. ethmoid foramen to post. ethmoid foramen (containing the post. ethmoidal artery)	**12** mm
From post. ethmoid foramen to optic foramen	**6** mm

(Ant. = anterior; Post. = posterior)

Information source: McCord, Tanenbaum, & Nunery, Oculoplastic Surgery, 3rd edition. Raven, 1995, p. 54.

SINUS DRAINAGE

The anterior and middle air cells of the ethmoid sinus, the maxillary sinus, and the frontal sinus all drain into the middle meatus. The posterior ethmoidal air cells drain into the superior meatus. The sphenoid sinus drains through the sphenoethmoid recess of each nasal fossa.

mnemonic The **3M** rule of sinus drainage:
Most sinuses drain into the **M**iddle **M**eatus.

The anterior and middle air cells of the ethmoid sinus, the maxillary sinus, and the frontal sinus all drain into the middle meatus.
Thus the **3M** rule: **M**ost sinuses drain into the **M**iddle **M**eatus.

mnemonic The **FRONT**al rule of sinus drainage:
The **FRONT**al sinus drains through the **FRONT**onasal duct to the **FRONT**al portion of the middle meatus.

mnemonic The **PES**ty rule of sinus drainage:
The drainage of the **P**osterior **E**thmoidal air cells is **PES**ty to remember since they drain to the **S**uperior meatus.

PE **P**osterior **E**thmoidal sinus drains to
S **S**uperior meatus

mnemonic Finally, the **Spheno**id sinus drains into the **Spheno**ethmoid recess of each nasal fossa!

PROPTOSIS

When the ophthalmologist encounters a patient with proptosis, there is a wide differential diagnosis that must be considered.

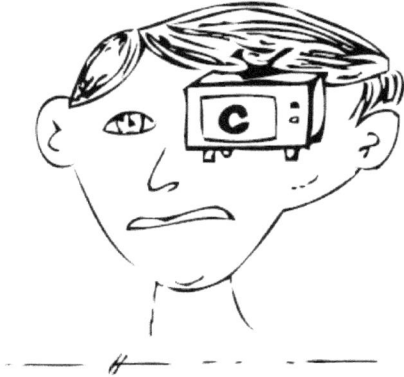

C TV IN the patient's eye with proptosis. (See Illustration)

Whenever you see a proptosis patient pretend you see **C** ("see") a **TV IN** the patient's eye (think of a color TV sitting in the patient's eye with proptosis).

Cyst (dermoid, epidermoid)
Thyroid disease (the most common cause of adult proptosis)
Vascular (carotid cavernous fistula or varix)
Inflammatory (pseudotumor or cellulitis)
NeoPla**SM** (e.g. rhabdomyosarcoma in child)
  -**P**rimary neoplasm
  -**S**econdary neoplasm
  -**M**etastasis (secondary orbital tumors are not synonymous metastases)

OR

An alternate mnemonic for proptosis is **VEIN**.*

Vascular
Endocrine
Inflammatory
Neoplasm

*VEIN mnemonic by permission of Lippincott, Williams & Wilkins/ Wolters Kluwer Health. From J. W. Henderson, in Essentials of Ophthalmology, G. B. Bartley, T. J. Liesegang (eds), J. B. Lippincott, Philadelphia, 1992, p. 205.

PSEUDOPROPTOSIS

Not all cases of "proptosis" are what they first appear. The eye, that at first glance appears proptotic, may actually be the normal eye in the case of pseudoproptosis.

You RIPa ME OFF

Remember – "You **RIPa ME OF**f!" for pseudoproptosis.
  (See Illustration)
Imagine that you send a proptosis patient to an Italian orbital specialist for surgery. He diagnoses the patient with pseudoproptosis and irately shouts "You **RIPa ME Of**f!" because he doesn't "C TV IN" the patient's eye. Now he's lost both a good TV and a good orbital case!

RIPa ME OFf!"

Retraction (lid)
Infantile glaucoma (buphthalmos)
Paresis of extraocular muscles
Myopia
Enophthalmos **O**ther eye
Facial asymmetry

| **THE EVALUATION OF ORBITAL DISORDERS** | When evaluating orbital disorders, there are several important components of the history and physical exam the ophthalmologist must remember. |

Remember - the Seven **P's** of orbital disorders.

Pain
Progression
Periorbital changes (lid lag/retraction or a conjunctival salmon patch)
Proptosis
Palpation
Pulsation
Plus auscultation

Mnemonic by permission of and modified from GB Krohel. GB Krohel in, Stewart W3, Chavis RM: Orbital disease: A Practical Approach. New York, Grune and Stratton, 1981.

AGE DISTRIBUTION OF CAPILLARY VERSUS CAVERNOUS HEMANGIOMA

Hemangiomas are benign vascular tumors or the orbit. Capillary hemangiomas are more common in pediatric patients while cavernous hemangiomas are more common in adults. To keep this age distribution straight in your mind:

Think - ca**P**illary hemangiomas occur in **P**ediatric patients.

STRAWBERRY NEVUS VERSUS PORT WINE STAIN

Two common cutaneous vascular tumors are the strawberry nevus of capillary hemangioma and the port wine stain of Sturge-Weber (which is so named because port wine is a very dark, deep red wine). The strawberry nevus of capillary hemangioma will blanch with pressure. The port wine stain of Sturge Weber will not blanch. To remember these distinctive clinical findings:

Think - **Port wine stain stays red when crushed** (pressure is applied), because the grapes that go to make it have already been crushed. When you crush them again with your finger (apply pressure), nothing happens. They've already been crushed.

While **Strawberry nevus blanches to white when crushed** (pressure is applied), because you are destroying the strawberries.

METASTATIC ORBITAL TUMORS IN CHILDREN

There are several tumors that commonly metastasize to the orbit in children. To remember these common metastatic orbital tumors:

Think – **NEW Leuk**ation for pediatric orbital metastases.

Imagine that the primary cancer cells of a pediatric orbital metastasis have found their **NEW Leuk**ation ("new location") in the orbit.

Neuroblastoma (the most common metastatic tumor to the orbit in kids)
Ewing's sarcoma
Wilms tumors
Leukemia (which, in almost **ALL** cases **A**cute **L**ymphocytic **L**eukemia)

Note: Granulocytic sarcoma and chloroma are other rare metastases.

PEDIATRIC ORBITAL NEOPLASMS & AGE-RELATED PROGNOSIS

For the pediatric orbital neoplasms primary neuroblastoma and langerhans cell histiocytosis (a.k.a. histiocytosis X), the age at diagnosis of the disease influences the survival rate.

Age at Diagnosis	Primary Neuroblastoma Survival	Langerhans Cell Histiocytosis Survival
< 1 year	90%	
> 1 year	10%	
< 2 years		50%
> 2 years		87%

NEuroblastoma is better in **Ne**wborns,
but **LA**ngherhans is better in **LA**ter age.

The prognosis for **NE**uroblastoma better in **Ne**wborns with a 90% survival for those diagnosed at less than one year of age.

However, the prognosis for **LA**ngherhans histiocytosis is better in **LA**ter age with an 87% survival at greater than 2 years, but only 50% less than 2 years.

Information sources: 1. American Academy of Ophthalmology Basic and Clinical Science Course: Orbit, Eyelids, and Lacrimal System. American Academy of Ophthalmology, 2008, p. 87.
2. American Academy of Ophthalmology Basic and Clinical Science Course: Orbit. American Academy of Ophthalmology, 1995, p. 45.
3. American Academy of Ophthalmology Basic and Clinical Science Course: Pediatrics. American Academy of Ophthalmology, 1990, p. 146.

ENLARGED EXTRAOCULAR MUSCLES ON NEURO-IMAGING

The most common cause of enlarged extraocular muscles on neuro-imaging is thyroid disease; however, there is a wide differential that must be considered as well.

Think - **FA₂TTI MPS** cause fatty muscles on neuro-imaging.

The most common cause of **FA₂TTI** ("fatty") muscles on neuro-imaging is thyroid eye disease, which is due to the increased production of the mucopolysaccharide **(MPS)** hyaluronic acid by fibroblasts.

Fistula (carotid cavernous fistula)
Amyloid, **A**cromegaly
Thyroid eye disease
Trichinosis
Infection (cellulitis & cavernous sinus thrombosis)

Metastasis (including Melanoma)
Pseudotumor of orbit
Sarcoid

PTOSIS

Ptosis can be of congenital, aponeurotic, mechanical, traumatic, myogenic, or neurogenic etiology. For this differential:

Think – "You **CLAM'T My Nerve** and caused this ptosis!"

Imagine you performed cataract surgery on a patient (cataract patients can complain of ptosis following cataract surgery). The irate patient returns with ptosis and shouts, "You **CLAM'T My Nerve** and caused this ptosis!"

"**CLAM'T My Nerve**" sounds like "clamped my nerve!"

Congenital
Levator **A**poneurotic
Mechanical
Traumatic
Myogenic
Nerve (neurogenic)

PSEUDOPTOSIS

Pseudoptosis may give the patient the appearance of ptosis, but this appearance is due to some other abnormality rather than true ptosis and hence will not fall under the "**CLAM'T My Nerve**" differential diagnosis for true ptosis.

"**Go through the motions**" for the differential of pseudoptosis.

This mnemonic is easy to remember if you "**go through the motions**." Start by pretending you are deep in thought and rest your hand against your brow. Then:

Pull down your eye brow	- Brow ptosis
Tent up your lid skin	- Dermatochalasis & Blepharochalasis
Push in your eye	- Enophthalmos
Look down	- Hypotropia
Pull up the lid of your other eye	- Contralateral lid retraction

DERMATOCHALASIS VERSUS BLEPHAROCHALASIS

Dermatochalasis is a redundancy of the skin of the eyelids, often associated with orbital fat protrusion that commonly occurs in middle-aged and older patients. Blepharochalasis is a rare familial condition characterized by idiopathic inflammatory edema of the lids. It occurs in young females.

Dermatochalasis occurs in the el**Der**ly & **D**otards and is seen **D**aily.

Blepharochalasis occurs in **Bl**oated **B**obbysoxers and is **B**arely seen. (The **Bl**oated **B**obbysoxers of **Bl**epharochalasis are young female patients)

NAMED LIGAMENTS OF THE LID	The named ligaments of the lid include the Superior Transverse Ligament of Whitnall (above) and the Suspensory Ligament of Lockwood (below). To remember the location of each:
	Remember - **L**ockwood's is **L**ower.
	Thus, the Suspensory Ligament of **L**ockwood is **L**ower, And Whitnall's "Superior" Transverse Ligament must be above.

THE LATERAL ORBITAL TUBERCLE	Many structures attach to the lateral orbital tubercle.
	Think - **L**ots of **L**'s attach to the **L**ateral orbital tubercle.
	Specifically there are 5 "**L**" structures and lots of accessory "**l**'s" along the way!
	Lateral canthal tendon, its **l**ower (posterior) crus
Lateral horn of the **l**evator aponeurosis
Lockwood's **l**igament
Lateral check **l**igament of the **l**ateral rectus
Lower (deep) **l**ayer of the orbital septum
	Information source: J. J. Dutton, Atlas of Clinical and Surgical Orbital Anatomy. W.B. Saunders, 1994, p. 3.

GLANDS OF MOLL	The glands of Moll are apocrine sweat glands that open at or near the lash follicles.
	Think of an (ugly **APE**-like) girl named **Moll**y with hyperhydrosis who **sweats**.
	This reminds you that glands of **Moll** are (**APE**-ocrine) **sweat** glands.
	Mnemonic by permission of and modified from Gary Brown, MD whose original mnemonic was "Think of a girl named **Moll**y with hyperhydrosis who **sweats**."

LYMPHATIC DRAINAGE OF THE MEDIAL EYELID	Lymphatic drainage of the lateral lids to the preauricular nodes is easily remembered because ophthalmologists are used to evaluating the patient for preauricular lymphadenopathy. Lymphatic drainage of the medial lids is a little more difficult, as the medial lids drain to the submandibular nodes (and checking for submandibular lymphadenopathy is not as routine).
	The **M**edial lid drains to sub**M**andibular node. (While the lateral lid drains to preauricular node.)

ANGULAR VEIN: LOCATION	The angular vein should be avoided during dacryocystorhinostomy. On average the angular vein is 8 mm medial to the medial canthal angle.
	8 mm medial = ∞ bleeding during DCR!
	The number "**8**" rotated 90 degrees resembles the symbol "∞" for infinity. Imagine that if you don't remember the location of the angular vein **8** mm medial to the medial canthal angle, you'll wind up with "∞ bleeding" during a DCR!

CLASSIFICATION OF ECTROPION

The differential diagnosis for ectropion includes congenital, involutional, cicatricial, inflammatory, mechanical, and paralytic etiologies. Keep this differential in mind by thinking:

A **CIC IMP** causes an ectropion. (See Illustration)

Imagine a **CIC IMP** ("a sick imp") hanging onto the lower lid and pulling it outward. The word "**CIC**" both sounds like the word "sick" and also reminds you of the etiology - **cic**atricial. An **IMP** is a mischievous child or little devil.

Congenital	**I**nflammatory
Involutional	**M**echanical
Cicatricial	**P**aralytic

Mnemonic by permission of and modified from Esther Rettig, MD.

TRICHIASIS VERSUS DISTICHIASIS

In trichiasis, the lashes arise from the correct position but the aberrant lashes come in contact with the globe. The lid margin may be in the normal position or inturned with entropion (e.g. chronic blepharoconjunctivitis or cicatricial conjunctivitis). Distichiasis is a relatively rare condition where an extra row of lashes arise from the meibomian glands. Remember distichiasis with this little rhyme:

Distichiasis is a **D**oubled lash row
Coming from the meibo.

OR

Simply remember *di-stichia* is from the Greek for two lines/rows

AND

In TR**I**chiasis the lashes **RI**se from the **RI**ght spot on the lid, but **trickle** onto the cornea.

TRICHIASIS: CRYOTHERAPY

When performing cryotherapy for trichiasis, the lid should be frozen to -20 degrees Celsius (with a thermocouple in the lid), then allowed to thaw completely, and then re-frozen to -20 degrees Celsius. This process requires a nitrous oxide probe to reach the necessary -20 degrees Celsius. A carbon dioxide probe will not freeze the eyelid to the required -20 degrees Celsius.

Remember – **The Rule of 2's** for cryotherapy for trichiasis.
Freeze to **-2**0 degrees Celsius **2** times.

AND

For cryotherapy for trichiasis you:
Need **N**itrous oxide and
Can't with **C**arbon dioxide.

EPIBLEPHARON VERSUS ENTROPION

In epiblepharon the pretarsal muscle and skin ride above the lid margin, causing the cilia to assume a vertical position. The "epi" in epiblepharon tells you there is overriding lid tissue. The lid margin is actually in the proper position and the problem stems from the overriding muscle and skin. In the case of entropion, on the other hand, the lid margin itself is turned inward. Epiblepharon usually occurs in children.

With e**P**iblepharon the lid margin is in the **P**roper **P**osition and it usually occurs in **P**ediatric **P**atients.

With e**N**tropion the lid margin is e**N**turned ("inturned").

TYPES OF EPICANTHUS

In epicanthus, there is a medial canthal fold of skin. In epicanthus tarsalis, the fold of tissue is most prominent in the upper lid. In epicanthus inversus, the fold of skin is greatest in the lower lid. In epicanthus palpebralis, there is an equal distributed fold of tissue in the upper and lower lid. (Epicanthus supraciliaris is a fourth type of epicanthus not mentioned in some texts.)

The **Tar-Pal-In(S)** of epicanthus.

Imagine that your patient with epicanthus has **Tar-Pal-In(S)** ("tarpaulins") covering the medial aspect of each eye!

Tar	**Tar**salis	**T**op web is larger
Pal	**Pal**pebralis	**Pa**rity between upper and lower folds
In	**In**versus	**In**ferior web is larger
(S)	**(S)**upraciliaris	**(S)**uperior web is larger – from the eyebrow to lacrimal sac

PERIPHERAL NERVE BLOCKS

Peripheral nerve blocks are useful for many ophthalmic procedures, especially laser skin resurfacing of the face. The supraorbital, infraorbital, and mental nerves all emerge through foramina that are at approximately the **vertical midpupil plane**, when the eye is in primary position.

Information source: Randle et al. J Dermatol Surg Oncol, 1992; 18: p. 231.

BLEPHAROPHIMOSIS

It is often difficult to remember all of the findings of blepharophimosis: Ptosis, Ectropion of the lateral lower lid, Telecanthus, and Epicanthus Inversus. To remember all of these findings:

Think – You **PEcTIE** the patient for the findings of blepharophimosis.

Imagine that you are operating on a patient to surgically correct their blepharophimosis. You **PEc** ("peck") away at the skin that **TIE**s the medial canthi together. (This is brief and inadequate description of a Y or V-plasty.) Thus, "You **PEcTIE** the patient for the findings of blepharophimosis."

Ptosis
Ectropion of the lateral lower lid
Telecanthus
Inversus type **E**picanthus

Mnemonic by permission of and modified from Bob Pashby, MD, FRCS(C), whose original mnemonic was "PETE."

GRAVE'S DISEASE: EPONYMS

There are numerous eponyms in Grave's disease. For all its worth, the three most common eponymous lid signs are:

Sign	Description
von Graefe's sign	Lid lag on downgaze
Dalrymple's sign	Eyelid retraction and stare
Stellwag's sign	Infrequent blinking

For von **Graefe**'s sign, imagine that a **Graefe** knife is held below the upper eyelids, and thus the upper lids are scared to move down with the globe on downgaze because they might be stabbed with the **Graefe** knife.
(Lid lag on downgaze is **von Graefe's** sign.)

For Dalrymple's sign, the word "**D-al-rymple**" sounds like "**all rumpled**." When there is eyelid retraction and stare, the eyeli**D** is **all rumpled** up and stays there!
(Eyelid retraction and stare is **Dalrymple's** sign.)

For Stellwag's sign, the word "Stellwag" sounds like "**stilled wag**." With infrequent blinking, the lids no longer wag, but instead stay still in a "**stilled wag**."
(Infrequent blinking is **Stellwag's** sign.)

LEVATOR APONEUROSIS AND LACRIMAL GLAND

It is the lateral horn of the levator aponeurosis that splits the lacrimal gland into its two lobes: the orbital and palpebral lobes of the lacrimal gland.

The **L**ateral horn of **L**evator splits the **L**acrimal gland into 2 **L**obes.

XANTHELASMA

Two-thirds of patients with xanthelasma do not have associated serum lipoprotein abnormalities. However young people with xanthelasma have a higher incidence of lipoprotein abnormalities, specifically homozygous type IIA hyperlipoproteinemia. To remember the latter point:

Uncross the lines in "**X**" in "**XA**nthalasma" and you will get the Roman numeral "**II**." Conveniently, the letter "**A**" is the next letter in the word **XA**nthalasma, reminding you that **XA**nthalasma is associated with type **IIA** hyperlipoproteinemia in younger patients.

Information source: D.S. Orentreich, F. Nesi & R. Lisma & M Levine (eds.), Dermatology of Eyelids in Smith's Ophthalmic Plastic & Reconstructive Surgery. Mosby, 1998, p. 501.

EPONYMS OF NASOLACRIMAL ANATOMY

We have all heard of the valve of Hasner, but it is not the only named structure in the nasolacrimal drainage system. Along this system, from proximal to distal, there are five structures with eponyms that may be quizzed. These are **M**aier's sinus (a dilation of common canaliculus where it enters the nasolacrimal sac), **R**osenmuller's valve, **K**rause's valve, **H**yrtl's spiral valve, **T**aillefer's valve, and **H**asner's valve.

Mayor Rose Krause Hurtles Tall Houses.

Mayor = Maier's sinus
Rose = Rosenmuller's valve
Krause = Krause's valve
Hurtles = Hyrtl's spiral valve
Tall = Taillefer's valve
Houses = Hasner's valve

Information source: McCord CD, Tanenbaum M, Nunery WR, eds, Oculoplastic Surgery, 3rd edition. Raven Press, 1995, p. 75.

NASOLACRIMAL INTUBATION: SOME HELPFUL MEASUREMENTS

For nasolacrimal intubation, it is useful to remember these helpful measurements. In infants, it is **20** millimeters from the punctum to the inferior meatus, and **20** millimeters from the external nare to the valve of Hasner. In adults the corresponding distances are **30** millimeters each

The 20/20 rule of nasolacrimal intubation in **infants.**

Reminding you in **infants** it's **20** mm from punctum to inferior meatus, and **20** mm from external nare to valve of Hasner.

The 30/30 rule of nasolacrimal intubation in **adults.**

Reminding you in **adults** it's **30** mm from punctum to inferior meatus, and **30** mm from external nare to valve of Hasner.

Mnemonic by permission of and modified from Joe Flanagan, MD, FACS.

NASOLACRIMAL DUCT: DRAINAGE

Anatomically, the nasolacrimal duct empties below the inferior turbinate. However, when creating an anastomosis during dacryocystorhinostomy surgery, an artificial opening is made whereby the nasolacrimal system now drains below the middle turbinate.

INitially the nasolacrimal duct drains under the **IN**ferior turbinate. But after the surgeon **MeDDLE**s it opens under the **MiDDLE**.

RELATIVE POSITION OF THE LACRIMAL PUNCTA

In normal eyelid anatomy, the lower (inferior) punctum is located more lateral than the superior punctum. The absence of this normal clinical finding may indicate lid disease.

The **L**ower puncta is more **L**ateral in the normal patient.

THE TESTS FOR TEAR PRODUCTION

The three most common tests for tear production are often quizzed and often confused (even by experienced attending physicians). The tests are as follows:

Test	Method	Measures
Basic secretion	Conjunctiva anesthetized Filter paper placed	Basic secretion ONLY from Krause & Wolfring glands
Schirmer I	No anesthesia Filter paper placed	Basic secretion AND reflex tearing
Schirmer II	Conjunctiva anesthetized Filter paper placed Middle turbinate is Tickled with a cotton swab	Reflex tearing ONLY

BAsiC: **A**nesthetize **C**onjunctiva
Schirmer **ON**e: **NO** anesthetic
Schirmer **TW**o: **T**ickle **T**urbinate, **W**ith anesthetic in conjunctiva

Thus, in a B**AsiC** Secretion test, you **A**nesthetize **C**onjunctiva and place the paper. The basic secretion test measures ONLY basic secretion from the glands of Krause & Wolfring.

In a Schirmer **ON**e test, there's **NO** anesthetic, you just place the paper – that's it! Thus a Schirmer **ON**e test measure basic secretion AND reflex tearing.

Finally, in a Schirmer **TW**o test you, **T**ickle **T**urbinate, **W**ith anesthetic in conjunctiva measuring reflex tearing ONLY.

POOR PROGNOSTIC FACTORS IN BELL'S PALSY

Bell's palsy, a lower motor neuron facial nerve palsy, can lead to corneal complications, especially in the presence of poor Bell's phenomenon, anesthesia of the cornea, or underlying dry eye.

The **BAD** syndrome causes corneal complications in Bell's palsy.

That is to say, Bell's palsy that results in corneal complications tends to be associated with the "**BAD** syndrome".

Bell's phenomenon (lack of)
Anesthesia of cornea
Dry eye

Mnemonic by permission of and modified from Thieme Medical Publishers. Mnemonic attributed to Pierre Guibor, MD by Robert E Levine, in "Eyelid Reanimation Surgery", Ch. 38, in Mark May's, The Facial Nerve, Theime Medical Publishers Inc, 1986.

ANCA AND WEGENER'S GRANULOMATOSIS

Anti-Neutrophil Cytoplasmic Antibody (ANCA) is measured by serum immunofluorescence and associated with systemic vasculitides, especially Wegener's granulomatosis. In particular, the cytoplasmic ANCA (c-ANCA) is more specific for Wegener's than the perinuclear ANCA (p-ANCA).

Imagine - Bald **C**harles **Anka** ("c-ANCA") who wears a **Wig** for his **Weg**ener's.

C-ANCA **C**orresponds better with Wegener's.

COMPLICATIONS OF LASER SKIN RESURFACING

The complications of laser skin resurfacing can be troublesome both for the patient and the ophthalmologist. These complications include redness, infection, pigmentation, and scarring.

When the laser **RIPS** the patient's skin you'll have problems!

Redness (persistent erythema)
Infection
Pigmentation (either hyper- or hypopigmentation)
Scarring

ERBIUM:YAG LASER VERSUS CO$_2$ LASER

The CO$_2$ laser ablates 80-150 microns per pass with increased thermal effect on tissue, thus coagulating tissue as it cuts. Therefore, the CO$_2$ laser can be used for laser incisional surgery. In contrast, the Erbium:YAG (Er:YAG) laser ablates only 10-40 microns per pulse, and causes a smaller zone of thermal damage. Although the smaller zone of thermal damage may be thought to cause fewer side effects, the thin ablation makes the Er:YAG laser unsuitable for incisional laser surgery.

The **C**0$_2$ laser **C**uts deeper and can be used for **C**utting because it **C**oagulates.

The **ER**:YAG has few**ER** side effects.

THE REVERSE TRENDELENBURG POSITION

In the reverse Trendelenburg position, the head of the bed is elevated. A slight reverse Trendelenburg position is often employed in orbital surgery in order to decrease the potential for bleeding during surgery.

Remember – **HURT** for Reverse Trendelenburg.

Imagine that you might **HURT** your reputation as an orbital surgeon if you forget that **H**ead **U**p is **R**everse **T**rendelenburg!

SUTURES: SQUARING KNOTS

To avoid fumbling while trying to square your instrument tie, just remember the following:

Put the needle driver in the middle and cross.

That is to say,
Put the needle driver in the middle (in between the suture ends)
And cross (cross the suture ends over the needle driver).

SUTURES: NON-ABSORBABLE

A variety of non-absorbable sutures is used in ophthalmic and oculoplastic surgery. These are known by both their generic and trade names – which can make remembering them difficult. To remember the non-absorbable sutures:

Remember – **The "L-N" Rule** of non-absorbable sutures.

The "L-N" Rule of non-absorbable sutures states:
If the last syllable of the suture name contains both an "L" and "N", chances are its very **Lo**N**g**-lasting (i.e. non-absorbable).
Examples: ny**LoN**, Pro**LeN**e, Mersi**LeN**e, Surgi**LeN**e, Surg**LoN**, Ethi**LoN**

AND

Remember – Non-absorbable sutures require **SNiPS**.

Non-absorbable sutures often must be removed requiring "**SNiPS**" (to be snipped out).

Silk
Nilon (nylon)
Prolene
Supramid

SCALPEL BLADES: CONFIGURATION

The different scalpel blades can be distinguished by their cutting edge, curvature, and angle. Commonly used blades in ophthalmology include the 15 blade, 11 blade, and #64 & # 66 Beaver blades. The 15 blade has a curved cutting edge and is commonly used in oculoplastics. The 11 blade is occasionally used in oculoplastics and is shaped like a triangle with a very sharp point. The #64 & #66 Beaver blades are commonly used to cut sclera. The #64 Beaver blade is a flat blade, while the #66 Beaver blade is bent at a sharp angle and therefore will lie flat on the sclera when held from above.

The Arabic numeral **5** has a curved surface.
Reminding you the **15** blade has a curved cutting edge.

The Arabic numeral **1** has a triangular apex.
So the two "1's" in the **11** blade remind you of its sharp triangular point.

Sixty-FOUR lies flat on the FLOOR.
Sixty-six is bent to pick up sticks.

Reminding you the #64 Beaver blade is flat.
And the #66 Beaver blade is the bent one (bent to more easily pick up sticks).

CLINICAL VARIANTS OF BASAL CELL CARCINOMA

Basal cell carcinoma is the most common eyelid malignancy. Classically, basal cell carcinoma is grossly described as a pearly, white nodule with telangiectatic vessels and an ulcerative central crater – a "rodent ulcer." But basal cell carcinoma does not always look like a "rodent ulcer," and can assume many different morphologies, including: pigmented, nodulo-ulcerative, nodular, cystic, superficial, and morpheaform (a.k.a. plaque-like).

"**SNiP'M Cyst**s" for the clinical variants of basal cell carcinoma.

"**SNiP'M Cyst**s" sounds like "Snip them cysts." Imagine that when you see basal cell carcinoma variants you need to "**SNiP'M Cyst**s" to get remove them – no matter what clinical variant you see.

Superficial
Nodular
i
Pigmented
Morpheaform
Cystic

OR

"**PUNC**h'**SM** off" for the clinical variants of basal cell carcinoma.

"**PUNC**h'**SM** off" sounds like "Punches them off" (i.e. to get rid of them). Think of a doctor who seeks out basal cell carcinoma variants and "**PUNC**h'**SM** off" to get rid of them.

Pigmented
Ulcerative (nodulo-**U**lcerative)
Nodular
Cystic
h
Superficial
Morpheaform

RECOGNITION OF CUTANEOUS MELANOMA

It may be difficult to distinguish a benign nevus from cutaneous melanoma. The **ABCDEF Rule** for cutaneous melanoma can help you recognize lesions suspicious for cutaneous melanoma.

The **ABCDEF Rule** for cutaneous melanoma.

Asymmetry
Border irregularity
Color variegation
Diameter of 6 mm or more
Evolutionary change (including enlargement, elevation, surrounding erythema, a hyperpigmented halo, size, symmetry, surface characteristics, pruritis, pain, bleeding, and tenderness.)
Funny looking lesions

Mnemonic by permission of and modified from Dermatology Online Journal as originally appeared by Brent P. Hazen, Ashish C. Bhatia, Tarif Zaim and Robert T. Brodell. The Clinical Diagnosis of Early Malignant Melanoma: Expansion of the ABCD Criteria to Improve Diagnostic Sensitivity, Dermatology Online Journal, 5(2):3, 1999.

NOTES & YOUR OWN MNEMONICS:

NOTES & YOUR OWN MNEMONICS:

VI. NEURO-OPHTHALMOLOGY

VISION LOSS: THE DIFFERENTIAL DIAGNOSIS

Vision loss may be due to a refractive error or a problem anywhere from the corneal tear film to the occiput. Functional visual loss, amblyopia, and nystagmus are other causes.

If you are sweating because you can't explain why someone can't see, **GO FAN** yourself OR go lie on the **SOFA**.

GO FAN	**SOFA**
Glasses to **O**cciput	**S**pectacles to **O**cciput
Functional	**F**unctional
Amblyopia	**A**mblyopia
Nystagmus	

Mnemonic by permission of and modified from Michael L. Slavin, MD as appeared in Review of Ophthalmology, 1995: Vol 2; p. 56-61.

IRIS ANATOMY

The iris is composed of four layers, the anterior pigment epithelium, anterior border layer, stroma, and the posterior pigment epithelium. The iris sphincter muscle is contained within the stroma, while the iris dilator muscle s contained within the anterior pigment epithelium. To remember the four layers of the iris and locations of these two muscles:

Think of a brown & yellow iris composed of **B.S.** & **PEe**.
(Pardon the language.)

Border (anterior **B**order layer)
Stroma
Pigment **E**pithelial layers
(the anterior and posterior **P**igment **E**pithelial layers)

Think of a strong **APE** (**A**nterior **P**igment **E**pithelium) pushing up and **dilating** the iris.

The dilator muscle is in the **A**nterior **P**igment **E**pithelium.

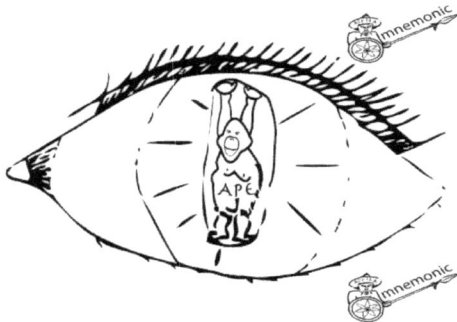

Iris **S**phincter is in **S**troma.

ANATOMY OF THE PUPILLOMOTOR FIBERS

The pupillomotor fibers travel with the 3rd cranial nerve. Thus both the pupil and ocular motility are often involved in an aneurysm of the posterior communicating artery that compresses the 3rd cranial nerve.

Remember – The **3 P's** of **P**upillomotor fibers:
Periphery, **P**ial and **P**osterior.

Pupillomotor fibers travel through the subarachnoid space in **P**eriphery of the 3rd cranial nerve.
They are nourished by the **P**ial vessels.
And they lie close to the **P**osterior communicating artery.

LOCATION OF THE CILIARY GANGLION

The ciliary ganglion is located between the lateral rectus and the optic nerve.

The words "ci**La**Ry gangli**ON**" tell you the ciliary ganglion's location.

LaR = **La**teral **R**ectus   gangli = ganglion   **ON** = **O**ptic **N**erve

So, the "ci**La**Ry gangli**ON**" is located between the **La**teral **R**ectus and the **O**ptic **N**erve.

LIGHT-NEAR DISSOCIATION

In medical school, we all learned that light-near dissociation is associated with the Argyll Robertson pupil of neurosyphilis. Now that we are studying neuro-ophthalmology, the differential widens to include at least these 11 disease processes: dorsal midbrain syndrome of Parinaud, diabetes, Adie's tonic pupil, Argyll Robertson pupil of neurosyphilis, amyloid, Dejerine-Sottas neuropathy (sometimes described as a subtype of Charcot-Marie-Tooth disease), Charcot-Marie-Tooth disease itself, aberrant regeneration, trauma, optic neuropathy, and finally severe retinopathy.

Wow! If there was ever a differential diagnosis that needed mnemonic, this is it!

Think - "**DASoCiATiON**" whenever you see light-near "dissociation."

Dorsal midbrain syndrome of Parinaud, **D**iabetes
Adie's, **A**rgyll Robertson, **A**myloid
Sottas (Dejerine-**So**ttas neuropathy), **S**evere retinopathy
Charcot-Marie-Tooth disease
Aberrant regeneration
Trauma
Optic **N**europathy

This is a difficult mnemonic so here are some clues for light-near **DASoCiATiON**.

Remember there are lots of **A**'s for light-near **DASoCiATiON**.
    Adie's
    Argyll Robertson
    Amyloid
    Aberrant regeneration
There are lots of eponyms for light-near **DASoCiATiON**.
    Dorsal midbrain syndrome of Parinaud
    Adie's
    Argyll Robertson
    Dejerine-**SO**ttas neuropathy
    Charcot-Marie-Tooth disease
There are two metabolic conditions for light-near **DASoCiATiON**.
    Diabetes
    Amyloid
There are two very rare conditions for light-near **DASoCiATiON**.
    Dejerine-**SO**ttas neuropathy
    Charcot-Marie-Tooth disease
Finally **A**berrant regeneration may result after **T**rauma.

ADIE'S TONIC PUPIL

Adie's tonic pupil is due to an idiopathic denervation of the parasympathetic pupillomotor fibers to the iris sphincter, resulting in a tonically dilated pupil. These pupils will constrict to dilute pilocarpine due to cholinergic denervation hypersensitivity. The condition is unilateral in 80% of cases.

"**Adie's**" sounds like "**eighties**" and the condition is unilateral **80%** of the time.

Information source: American Academy of Ophthalmology Basic and Clinical Science Course: Neuro-Ophthalmology. American Academy of Ophthalmology, 2008, p. 266.

FINDINGS IN ARGYLL ROBERTSON PUPILS

Argyll Robertson pupils are miotic pupils associated with neurosyphilis. They can be irregular and will constrict with accommodative effort but have lost the ability to constrict to light. To remember the findings of the Argyll Robertson pupil:

Think of the **5 L's** of the Argy**LL** Robertson pupil.

Loss of **L**ight reflex (accommodative reflex preserved)
Little, **L**opsided pupils (miotic, irregular pupils)
Associated with the **L**ues (syphilis)

Mnemonic by permission of and modified from Steve Kraft, MD.

A RELATIVE AFFERENT PUPILLARY DEFECT WITHOUT A VISUAL FIELD DEFECT

There is one neuro-ophthalmologic lesion that can cause a relative afferent pupillary defect (RAPD) *without* a corresponding visual field defect. This lesion occurs at the Brachium of the Superior Colliculus (abbreviated here **B.Sc.**), and causes a *contralateral* RAPD *without* a visual field defect.

You deserve a **B.Sc.** in neuro-ophthalmology in you remember that:
A lesion of the **B.Sc.** (**B**rachium of the **S**uperior **C**olliculus) causes a *contralateral* RAPD *without* a visual field defect!

This phenomenon occurs because the pupillary fibers exit the visual pathway prior to the lateral geniculate body, and enter the brain stem at the **B.Sc.** (**B**rachium of the **S**uperior **C**olliculus). A lesion at the **B.Sc.** (**B**rachium of the **S**uperior **C**olliculus) therefore results in a *contralateral* APD *without* a visual field defect.

THE MIOTIC PUPIL

When a patient presents to the clinic with miotic pupils:

Remember – The **A's & P's** of sm**A**ll **P**upils.

Agents (**P**ilocarpine, **P**esticides & o**P**iates)
Adie's (causes chronic miosis)
Age (more common in babies and the elderly)
Argyll Robertson
Aches (pain)
Accommodative spasm
Attention (drowsiness)
Pontine infarcts with **P**inpoint **P**upils
and finally Horner's Syndrome (sorry, no "A" or "P" here)

PUPIL SIZE IN DRUG OVERDOSE	Morphine overdose causes miotic pupils. Amphetamine overdose causes dilated pupils.

Morphine makes the pupils **Morph-IN** (i.e. constrict).

AMPhetamine makes the pupils **AMP**lify (i.e. dilate).

FINDINGS IN PARINAUD'S DORSAL MIDBRAIN SYNDROME	Perinaud's dorsal midbrain syndrome presents with a constellation of eye findings due to compression of the dorsal midbrain (specifically the superior colliculus within the mesencephalon). Causes include pinealoma, multiple sclerosis, and stroke. To remember the finding of Perinaud's dorsal midbrain syndrome:

Remember – The word "**PARINAUD'S**" spells out the finding of Perinaud's dorsal midbrain syndrome.

Papilledema
Accommodative insufficiency
Retraction of lids (Collier's sign)
Insufficient convergence
Nystagmus (convergence-retraction **N**ystagmus)
Aqueductal stenosis (in some cases)
Upgaze deficit
Dissociation (light-near **D**issociation of the pupil)
Skew deviation

Mnemonic by permission of and modified from Ravi Krishnan, MD.

HORNER'S SYNDROME	Patients with Horner's syndrome present with ptosis, miosis, and anhydrosis as well as absence of the spinociliary reflex (pupil dilation with stimulation to cranial nerve V or spinal nerves C2 or C3). To remember the four signs of Horner's syndrome:

Remember - The 4 **S**'s of Horner's syndrome.

Slight ptosis
Small pupil (miosis)
Sweating absent
Spinociliary reflex absent

Mnemonic modified from John J. Shipman, Mnemonics and Tactics in Surgery and Medicine, 2nd edition. Year Book Medical Publishers, 1984, p. 181.

USE OF PAREDRINE IN HORNER'S SYNDROME	When using hydroxyamphetamine (trade name Paredrine) in the testing of Horner's syndrome, if the Horner's pupil dilates more than the normal pupil, the lesion is probably in the preganglionic neuron. On the other hand, if the Paredrine dilates the normal pupil more than the Horner's pupil and the anisocoria increases by at least 0.8 mm (i.e. the anisocoria gets worse with Paredrine), then the lesion could be considered postganglionic with reasonable certainty.

A **PaRE**drine **Re**sponse = A **PRE**-ganglionic **PR**ocess.

This reminds you that dilation of a Horner's pupil with Paredrine indicates damage to a preganglionic neuron in Horner's syndrome.

SOME OPTIC NERVE NUMBERS

You can remember many aspects of optic nerve anatomy and function with this mnemonic.

Approximations of 1 and 5 for the optic nerve.

There are approximately **1.5** million optic nerve fibers per nerve (actually 1.4 million).

The optic nerve head is **1.5** mm in diameter (slightly larger in the vertical dimension).

The blind spot is **15** degrees temporal to fixation.

The blind spot occupies about **5** degrees of visual space (slightly larger in the vertical dimension).

The length of the intraocular optic nerve is **1** squared mm = 1 mm.

The length of the intraorbital optic nerve is **5** squared mm = 25 mm.

Information source: American Academy of Ophthalmology Basic and Clinical Science Course: Neuro-Ophthalmology. American Academy of Ophthalmology, 2008, p. 25.

THE LENGTH OF OPTIC NERVE SEGMENTS

The optic nerve is divided into the intraocular, intraorbital, intracanalicular, and intracranial portions; and the length of each of these segments is often quizzed. The shortest portion, of course, is the intraocular portion which is only 1mm long. It is followed by the longest portion, the intraorbital which is 25mm long. Then there is the intracanalicular at 9mm and the intracranial at 16mm. Building off of the **Approximations of 1 and 5** for the optic nerve above, you can remember these lengths with this little rhyme:

Take the square of 1, 5, 3, & 4
From the globe to the core!

The "core" is the intracranial space as you proceed from the more superficial globe to the deeper intracranial space (i.e. the "core").

1 squared = 1 mm, the intraocular portion of the optic nerve
5 squared = 25 mm, the intraorbital portion of the optic nerve
3 squared = 9 mm, the intracanalicular portion of the optic nerve
4 squared = 16 mm, the intracranial portion of the optic nerve

Note: If have trouble recalling the numbers and rhyme, think that the sequence goes short, long, short, long. The shortest segment (intraocular) is paired with the longest segment (intraorbital).

Information source: American Academy of Ophthalmology Basic and Clinical Science Course: Neuro-Ophthalmology. American Academy of Ophthalmology, 2008, p. 25.

AUTONOMIC INNERVATION OF THE EYE

The parasympathetic fibers innervate the ciliary body and iris sphincter through the short ciliary nerves. In contrast, the sympathetic fibers innervate the iris dilator and ciliary body through the long ciliary and nasociliary nerves. To remember these innervation patterns:

Think of a **PAIR of SHORTS** for the short ciliary nerves.

The "**PAIR of SHORTS**" reminds you that the **PAIR**-asympathetic nerves travel through the **SHORT** ciliary nerves.

Then by exclusion, the sympathetic fibers must travel through the long ciliary (and nasociliary) nerves.

OPTIC DISC SWELLING:
DIFFERENTIAL DIAGNOSIS
When you look in and see optic disc swelling on exam, first, don't panic. Second, take a deep breath. And third, go through the differential diagnosis for optic disc swelling in the eye with **the 12 "I's"** of disc edema listed below.

Remember – Optic disc edema is a big "**I**" ("eye") problem,
So think of **the 12 "I's"** of optic disc edema.

Imitators (hyperopia, drusen, & myelinated nerve fiber layer)
Increased pressure (systemic hypertension, increased ICP, & orbital tumor)
Inversed pressure (e.g. hypotony)

Ischemia
Inflammation (papillitis, papillophlebitis, neuroretinitis, & uveitis)
Infiltration (glioma, meningioma, lymphoma, & metastases)

Insulin-related (e.g. diabetic papillopathy)
Inherited (e.g. Leber's hereditary optic neuropathy)
Insufficiencies (anemia, & hypoxemia)

Intoxications (e.g. methanol intoxication)
Irvine-Gass Syndrome
Inferior lesions (spinal lesions occasionally cause papilledema)

STELLATE
NEURORETINITIS
The most common cause of stellate neuroretinitis is cat scratch disease caused by the bacterium *Bartonella henselae*. However, a variety of disease processes may cause stellate neuroretinitis (or a very similar clinical picture).

Remember - "**STeLLA HaD** neuroretinitis" for stellate neuroretinitis.

Scratch Disease (cat **S**cratch disease), **S**yphilis
Toxoplasmosis, **T**oxocariasis
e
Lyme, **L**eptospirosis
Leber's (a.k.a. **L**eber's idiopathic stellate neuroretinitis)
ARN (Acute retinal necrosis)

HSV, **H**epatitis B & **H**istoplasmosis
a
DUSN (Diffuse Unilateral Subacute Neuroretinitis)

Note: Stellate neuroretinitis is rarely if ever associated with multiple sclerosis.

Information source: C. H. Smith, Miller and Newman (eds), Optic Neuritis in Walsh and Hoyt's Clinical Neuro-Ophthalmology 6th edition, Vol 1. Lippincott Williams & Wilkins, 2004, pp. 333-336.

LEBER'S HEREDITARY OPTIC NEUROPATHY (LHON)

The most common primary mutations in LHON are 11778 (Wallace), 3460, 14484, and 15257. (I have no mnemonic for these mutations other than three of them are 5 digits long and begin with 1. Also 1 + 7 = 8, 4 + 4 = 8, and 2 + 5 = 7). Thus the 11778 mutation is the one with 1's and 7's that = 8, the 14484 mutation is the one with 4's and 4's that = 8, and the 15257 mutation is the one with 2's and 5's that = 7.

Of note, the 14484 mutation of LHON has the best prognosis for visual improvement.

The 1**44**8**4** mutation for LHON contains a lot of **FOURS** because these patients with LHON are more **FOUR**tunate than others.

Information source: DR Johns et al. Arch Ophthalmol, 1993; 111: pp. 495-498.

WALLERIAN DEGENERATION

If the axons of nerves are permanently disrupted, over a period of days to weeks the distal part of the axon degenerates. This is called Wallerian or ascending (anterograde) degeneration. For Wallerian degeneration:

Think of **climbing** and ascending a **wall**.

Climbing a **wall** reminds you that Wallerian ("**wall**") degeneration is an ascending ("**climbing**") degeneration of the nerve axors.

SELLAR & SUPRASELLAR LESIONS

A variety of sellar and suprasellar lesions may affect the visual system by compression of the optic chiasm or the adjacent optic tracts or nerves as they intersect the chiasm. Most can remember pituitary lesions and craniopharyngioma, but the differential is broader than these two lesions. For this differential:

Remember - **CAMPDOGI** for sellar and suprasellar lesions.

Craniopharyngioma
Aneurysm, **A**rachnoid cyst
Meningioma
Pituitary lesion
Dermoid
Optic nerve glioma
Glioma of the hypothalamus
Infiltration (TB & sarcoid)

Mnemonic origin unknown. I have seen this popular mnemonic used by many neurosurgeons and neuroradiologists.

DRUGS USED IN MYASTHENIA GRAVIS	Edrophonium (trade name Tensilon) is a short-acting cholinesterase inhibitor used in the testing for myasthenia gravis with the "Tensilon Test." Pyridostigmine (trade name Mestinon) is a long-acting cholinesterase inhibitor used in the treatment of myasthenia gravis.

Tense Eddie is used for the myasthenia "Tensilon Test."

Tense = Tensilon   **Eddie** = Edrophonium

Messy Pyramids are used to treat myasthenia

Messy = Mestinon   **Pyramids** = Pyridostigmine

CROSSING OF THE TROCHLEAR NERVES	The trochlear nerves exit the brainstem dorsally and decussate at the anterior medullary velum. The most likely cause of bilateral superior oblique palsies following trauma such as a motor vehicle accident is a lesion at the anterior medullary velum where the trochlear nerves decussate.

Remember – Bilateral superior oblique palsies:
Commonly follow an **MVA** with a lesion of the **AMV**.

That is to say, bilateral superior oblique palsies commonly follow trauma such as an **MVA** (**M**otor **V**ehicle **A**ccident) with trauma to the **AMV** (**A**nterior **M**edullary **V**ellum) where the trochlear nerves decussate.

Now take your pen and make the last stroke of the letter "M" bold, in both **AMV** and **MVA**. Make the letter "V" in both bold as well. (See Illustration)

$$MVA \Rightarrow AMV$$

This gives you two roman numerals "**IV**," reminding you of bilateral superior oblique palsies (CN **IV** palsies) following an **MVA** with trauma to the **AMV**.

CENTRAL CAUDAL NUCLEUS	Much of the motor innervations to the extraocular and facial muscles is unilateral in nature. One exception to this is the central caudal nucleus which sends projections to both levators. To remember that the central caudal nucleus sends fibers to both levators, remember this little rhyme:

**The nucleus central caudal
Makes both lids upward wobble.**

EYE MOVEMENT

1) **Saccades** – Both the superior colliculus and frontal eye fields (in Brodmann area 8) contribute to the production of saccades.

BOTH the **S**uperior **C**olliculus contributes to **SaC**cades,
AND the Frontal eye fields (Brodmann area **8**) contribute to Sa∞ades.

First, the "**S**" and "**C**" of **S**uperior **C**olliculus and the "**S**" and "**C**" of **SaC**cades go together.

Second, take the number "**8**" from Brodmann area **8** and lay it on its side. Now it resembles two letters "c" touching each other, and there are two "c's" in the word Sa∞ade. Also, if you draw a pair of eyes in the circles formed, it reminds you that Brodmann area 8 is in the frontal lobes, which subserve eye movement. (See Illustration)

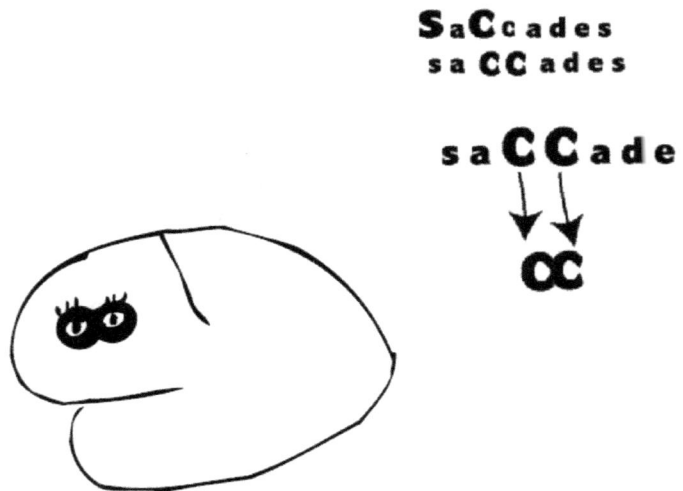

$$\text{Sa\textbf{C}cades}$$
$$\text{sa\textbf{CC}ades}$$

$$\text{sa\,}\textbf{C}\textbf{C}\text{\,ade}$$
$$\downarrow\downarrow$$
$$\textbf{CC}$$

Graphic above adapted by permission of Stephen Goldberg, MD. From Stephen Goldberg, MD, Clinical Neuroanatomy Made Ridiculously Simple. MedMaster, Inc., Miami: 2007, figure 52, p 72.

2) **Saccade Regulation** – The nucleus raphe interpositus contains pause cells (a.k.a omnipause cells) which regulate saccades by gating the activity of burst neurons.

Pause cells are located in the nucleus raphe Inter"**pause**"itus.

3) **Smooth Pursuit** – The parieto-occipital-temporal junction (abbreviated here as **POT** junction) is involved in the generation of smooth pursuit eye movements.

The **POT** junction generates smooth **P**ursui**T** eye movements.

PARKS THREE-STEP TEST

The Parks three-step test is used in patients with hyperdeviation to determine which of the vertically acting muscles is palsied. The eye with the hyperdeviation is documented in primary position, right & left gaze, and right & left head tilt. The test and its outcomes may seem complicated or confusing to all but the strabismologist. For practical clinical purposes however, it helps to identify if a patient has a superior oblique palsy, and which side is affected.

A **Right** superior oblique palsy is **Right, Left, Right**.

On the Parks three-step test, the patient with a **Right** superior oblique palsy has a **Right** hyperdeviation that is worse in **Left** gaze and **Right** head tilt.

A **Left** superior oblique palsy is **Left, Right, Left**.

On the Parks three-step test, the patient with a **Left** superior oblique palsy has a **Left** hyperdeviation that is worse in **Right** gaze and **Left** head tilt.

Mnemonic origin unknown.

CHARACTERISTICS OF BILATERAL TROCHLEAR NERVE PALSY

In any patient with unilateral trochlear nerve palsy, bilateral trochlear nerve palsies should be expected. The characteristics of bilateral trochlear nerve palsy include positive bilateral Bielchowsky head tilt tests, V pattern esotropia on downgaze, and excyclotorsion greater than 10 degrees.

Bilateral trochlear nerve palsies can be a **BiT VEX**ing.

BiT	= **Bi**lateral **T**rochlear palsies have **Bi**lateral **Ti**lts with
VE	= **V** pattern **E**sotropia on downgaze and
X	= **EX**cylotorsion of greater than **X** (Roman numeral for 10) degrees.

DOWNBEAT NYSTAGMUS IN THE PRIMARY POSITION

Lesions of the craniocervical junction such as Arnold-Chiari malformation, cerebellar atrophy, basilar invagination, and multiple sclerosis can cause downbeat nystagmus.

When you see a patient with downbeat nystagmus, **think low**.

The lesions which cause downbeat nystagmus occur **low**-down in the brain at the craniocervical junction such as Arnold-Chiari malformation, cerebellar atrophy, basilar invagination, and multiple sclerosis

Mnemonic origin unknown. Special thanks to J. Lawton Smith, MD for mentioning it.

| **PSEUDO-SIXTH NERVE PALSIES** | Sixth nerve palsy leaves the patient unable to abduct the eye due to weakness of the lateral rectus muscle. However, there are other causes apparent weakness of abduction that may look like sixth nerve palsies, the "pseudo-sixth nerve palsies." |

Don't **MiSiDENTiF**y pseudo-sixth nerve palsies.

Myasthenia
i
Spasm of the near reflex
i
Duane's type I
Esophoria **N**o longer compensated (a.k.a. decompensated esophoria)
Thyroid
i
Fracture of the medial wall

If you're still having trouble identifying the cause of a PSEUDO-sixth, reverse the letters of the word "PSEUDO" to yield "o**DUESP**"-sixth and take two letters at a time!

o
DUane's type I
ESophoria decompensated
SPasm of the near reflex

| **FASCICULAR SYNDROMES OF CRANIAL NERVE 3** | 1) The third nerve nucleus is located in the midbrain. The named CN 3 fascicular syndromes are **N**othnagel's, **B**enedikt's, **W**eber's, and **C**laude's syndromes. Remember these fascicular syndromes of CN 3 with this little rhyme: |

N, B, W, C,
Fascicular syndromes of CN 3.

That is **N**othnagel's, **B**enedikt's, **W**eber's, and **C**laude's syndromes are the fascicular syndromes of CN 3.

Think of this little rhyme and these important **NB** eponyms every time you go to the **WC** (water closet or bathroom) and you will quickly have the fascicular syndromes of CN 3 memorized!

2) Nothnagel's syndrome consists of CN 3 palsy and ipsilateral cerebellar ataxia.

Nothnag**EL** affects the cereb**EL**.

Nothnag**EL** rhymes with cereb**EL**. Thus Nothnag**EL**'s syndrome gives CN 3 palsy with ipsilateral cereb**EL**-lar ataxia.

3) Benedikt's syndrome involves the red nucleus, and therefore a contralateral "rubral tremor" is one of the manifestations.

mnemonic

Benedict Arnold **trembled** when the **Red**coats were coming.

Benedict = Benedikt's syndrome
trembled + **Red**coats = Rubral tremor

Thus Benedikt's syndrome consists of CN 3 palsy with contralateral rubral tremor.

4) Weber's syndrome consists of an ipsilateral third nerve palsy and contralateral spastic paralysis.

mnemonic

For **WEB**er's syndrome, imagine a person with a **WEB** over half his face restricting his CN 3 function, and another **WEB** binding his contralateral arm and leg causing hemiparesis (contralateral spastic paralysis).

NYSTAGMUS THAT STAYS HORIZONTAL IN VERTICAL GAZE

Three types of nystagmus usually remain horizontal uniplanar, even on vertical gaze. These three are congenital nystagmus, periodic alternating nystagmus, and peripheral vestibular nystagmus. To remember these three types use this little rhyme:

mnemonic

**Congenital, Per-AN and Per-Ve
Nystag Horizontally!**

This reminds you that **Congenital**, **Per**iodic **A**lter**N**ating, and **Per**ipheral **V**estibular Nystagmus all remain horizontal uniplanar, even on vertical gaze.

THE MYOCLONIC TRIANGLE

Oculopalatal myoclonus is pendular nystagmus with synchronous movement of the palate. This condition is due to a lesion in the area of the "myoclonic triangle", which connects the following areas: 1. the ipsilateral red nucleus 2. the ipsilateral olivary nucleus and 3. the contralateral dentate nucleus. To remember the structures of the "myoclonic triangle,"

IPSILATERAL

CONTRALATERAL

mnemonic

Picture - **Ruby red lips** trying to drink from a **triangle-shaped martini glass** with an **olive** on the bottom. Before the lips touch the glass, the person's **dentures** fall out on the **other side**.
(See Illustration)
The **triangle-shaped martini glass** represents the "myoclonic triangle" which is the area responsible for oculopalatal myoclonus.

The rest of the picture represents its three components.
The **Ruby red lips** represent the ipsilateral red nucleus.
The **olive** represents the ipsilateral olivary nucleus.
And the **dentures** that fall out on the **other side** represent the contralateral dentate nucleus.

FASCICULAR SYNDROMES OF CRANIAL NERVE 6

1) The sixth nerve nucleus is located in the pons. The CN VI fascicular lesions are Foville's syndrome (dorsal pontine lesion) and Millard-Gubler syndrome (ventral pontine lesion). A third type of sixth nerve palsy is Raymond's syndrome, which consists of abducens palsy plus contralateral hemiplegia.

FO**VI**LLE'S, **MI**LLARD-GUBLER, and RA**Y**MOND'S are all CN **VI** fascicular lesions.

The CN **VI** lesions are easy to recognize if you capitalize the eponym, and look for a Roman numeral **VI** in the letters. FO"**VI**"LLE'S is the easiest to recognize. For **MI**LLARD-GUBLER you have to extract the Roman numeral "**V**" from the "M." For RA**Y**MOND'S the letter "**Y**" can be thought of as a Roman numeral "**V**" on top of an "**I**."

OR

FOVI**66**E'S and MI**66**ARD-GUBLER are two CN **6** fascicular lesions.

You can add circles to the bottom of the double "L's" in FOVILLE'S and MILLARD-GUBLER to convert the double "L's" to double "6's." This mnemonic does not include Raymond's syndrome.

2) Generally speaking the brainstem is organized so that the spinothalamic fibers (the pain/sensation fibers) run posterolaterally, and the corticospinal fibers (the motor fibers) run along the anterior brainstem. By following this logic, you can guess many of the manifestations of brainstem disease depending on whether they affect the posterior or the anterior brainstem.

Remember - **Post**men are **Sens**itive people, and **Ant**s crawl to **Move**.

Thus the **Post**erior brainstem (in general) is **Sens**ation & pain while the **Ant**erior is **Move**ment/motor.

3) As mentioned before, fascicular CN 6 lesions include Foville's syndrome (a dorsal pontine lesion) and Millard-Gubler syndrome (a ventral pontine lesion). To remember that Foville's involves the dorsal pons:

Think - F**O**ville's is a lesion of the d**O**rsal or p**O**sterior p**O**ns. Therefore, by exclusion Millard-Gubler is a lesion of the ventral pons.

The "O" in F**O**ville's should remind you of the d**O**rsal or p**O**sterior p**O**ns.

POSTERIOR INTERNUCLEAR OPHTHALMOPLEGIA (INO) WITH PRESERVED CONVERGENCE

According to Cogan's designation, an anterior INO often impairs the convergence mechanism since it is usually due to a mesencephalic lesion which involves the convergence center. However, with a posterior INO convergence remains intact.

Remember - In a **P**osterior INO convergence is **P**reserved.

CALORIC NYSTAGMUS

Caloric nystagmus utilizes the vestibulo-oculuar reflex to test for brain stem death. You can elicit caloric nystagmus in a patient by irrigating cold or warm water in the patient's external auditory meatus either unilaterally or bilaterally. The direction of the nystagmus is determined in the United States by the fast phase of the movement. For a patient with an intact, functioning brain stem the following apply:

1) In UNILATERAL caloric stimulation, the direction of the fast (jerk) phase is remembered by the mnemonic:

The **COWS** Rule for UNILATERAL caloric stimulation.
Cold **O**pposite, **W**arm **S**ame.

Cold water irrigation results in the fast phase of nystagmus to the **O**pposite side as the ear irrigated;
Warm water irrigation results in the fast phase of nystagmus to the **S**ame side as the ear irrigated.

Mnemonic origin unknown. Special thanks to J. Lawton Smith, MD for mentioning it.

2) For the COMATOSE patient, caloric stimulation will reveal no fast phase. Instead, the patient will have only a tonic deviation of the eyes towards the side of cold water. To remember the absence of the fast phase in comatose patients:

Think - Comatose patients don't move fast anymore.

Thus caloric stimulation in the comatose patient will only lead to a slow, tonic deviation of the eye toward the side of cold water stimulation. Notice that this still obeys the COWS Rule, as the COWS rule refers to the fast of phase of nystagmus, which is notably absent in the comatose patient.

3) In BILATERAL simultaneous caloric stimulation, the direction of the fast (jerk) phase is vertical and can be remembered by this mnemonic:

The **CUWD** Rule of BILATERAL caloric stimulation:
Cold **U**p, **W**arm **D**own.

Cold water irrigation in both ears results in an **U**pward fast phase nystagmus.
Warm water irrigation in both ears results in a **D**ownward fast phase nystagmus.

Mnemonic origin unknown.

OR

Remember - If both your feet are on fire you'll look down quickly!

Both feet on fire reminds you of bilateral warm stimulation.
Looking down quickly reminds you of downward fast phase nystagmus.

Mnemonic by permission of and modified from J. A. Leavitt, MD.

INTERNUCLEAR OPHTHALMOPLEGIA	Internuclear ophthalmoplegia (INO) results from a lesion of the ipsilateral medial longitudinal fasciculus. It produces a pseudo-medial rectus weakness on adduction. Abducting nystagmus is characteristic, but need not be present. To remember the appearance and anatomy of an INO:

mnemonic Think – **INO** stands for **I**nsufficient **N**asal muscle **O**utput with an **I**psilateral MLF lesion.

THE "HALF" IN THE ONE-AND-A-HALF SYNDROME	The one-and-a-half syndrome is caused by a lesion to the MLF associated with an ipsilateral lesion to either the paramedian pontine reticular formation (PPRF) or a lesion to the abducens nucleus. The "one" in one-and-a-half syndrome is due to the PPRF or abducens lesion causing an ipsilateral gaze palsy. In other words, a lesion to the left PPRF or abducens nucleus prevents both eyes from moving to the left. The "half" in one-and-a-half syndrome is due to the MLF lesion which causes an adduction deficit in the ipsilateral eye. In other words, the associated lesion to the left MLF prevents the left eye from adducting (moving to the right).

mnemonic The M**LF** lesion accounts for the "ha**LF**" in one-and-a-ha**LF** syndrome.

PARANEOPLASTIC CEREBELLAR DEGENERATION AND OPSOCLONUS: ANTI-NEURONAL ANTIBODIES	Patients with paraneoplastic cerebellar degeneration may present to the ophthalmologist for evaluation of downbeat nystagmus, diplopia, or opsoclonus. Opsoclonus in an adult is often due to small cell carcinoma of the lung, but breast cancer has also been reported. The antineuronal antibody tests and their neoplastic associations of relevance to ophthalmology are as follows:

mnemonic Anti-**Y**o Ab is associated with g**Y**necologic cancer.
Anti-**H**u Ab is associated with c**H**est (lung) cancer.
Anti-**R**i Ab is associated with b**R**east cancer.

Note: The anti-**Y**o antibody (a.k.a. anti-Purkinje cell antibody) is so specific for g**Y**necological tumors that, in some cases, surgical exploration of the pelvis is recommended in women with the antibody, even if other diagnostic studies do not suggest a malignancy.

Information Sources: 1. Hetzel DJ, Mayo Clin Proc, 1990: 65, p1558.
2. Furneaux HM, NEJM, 1990: 322, p1844.
3. Luque FA, Ann Neurol, 1991: 29, p 241.

VISUAL PATHWAY ANATOMY	Tracing the visual pathway from the retina to the occipital cortex can be useful in diagnosing the cause of visual field defects. You can trace the pathway of the inferior retinal fibers using the mnemonic below.

mnemonic The many **L**'s the **L**ower retinal fibers.

There are Lot's of **L**'s: The **L**ower (inferior) retinal fibers are found in the **L**ateral part of the **L**ateral geniculate body, and after synapsing proceed to the **L**ingual gyrus at the **L**ower **L**ip of the calcarine fissure.

CEREBRAL BLOOD FLOW

Cerebral blood flow comes 70% from the carotid arteries and a 70% obstruction of the carotids is clinically significant.

Remember - **The 70/70 Rule** of cerebral blood flow.

70% of cerebral blood flow is from the carotid arteries.
A **70%** obstruction is needed before a stenosis is considered hemodynamically significant.

HEMIPARESIS AND LESIONS OF THE OPTIC RADIATIONS

Small infarcts of the inferior optic radiations may result in contralateral superior hemianopsia and contralateral hemiplegia, due to adjacent internal capsule involvement.

Remember - The **IN**ferior optic radiat"**IN**"s lie close to the **IN**ternal capsule.

HOMONYMOUS HEMIANOPSIA AND THE OKN RESPONSE

Cogan observed that in the presence of a homonymous hemianopsia, OKN asymmetry suggests a deep parietal lesion. OKN asymmetry is defined as a decreased OKN response when the drum is rotated toward the side of the lesion. Deep parietal lesions disrupt optomotor fibers intended for the ipsilateral pons, and thereby disrupt smooth pursuit to the ipsilateral side. Patients with homonymous hemianopsia due to an optic tract, temporal lobe, or purely occipital lobe lesion usually have symmetric OKN responses to both sides.

Imagine taking your **OKAP** exam in a **H**emi-**H**ome constructed of half an OKN drum. (See Illustration)

The **H**emi-**H**ome represents **H**omonymous **H**emianopsia.
OKAP and the OKN drum reminds you that **OK**N **A**symmetry suggests a **P**arietal lesion.

AND

The OKN response is **D**ecreased when the **D**rum is **D**irected at the **D**eep parietal lesion.

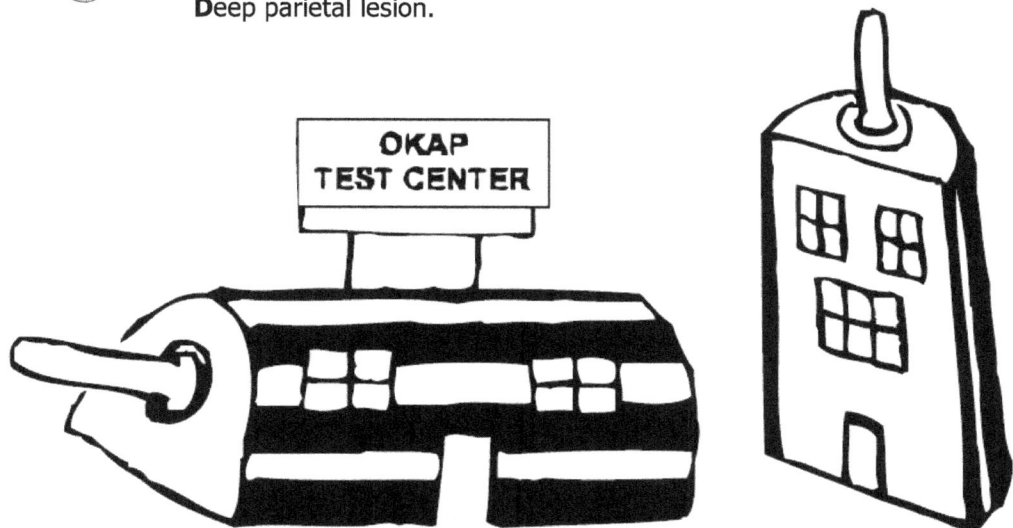

OKAP TEST CENTER

INTERFERON	Interferon beta 1a and beta 1b may be used in the treatment of multiple sclerosis. To remember the trade name of each drug:

Interferon beta 1**A** is **A**vonex (trade name)
Interferon beta 1**B** is **B**etaseron (trade name)

THE "NEUROLOGIC CAUSES" OF CONSTRICTED VISUAL FIELDS	When considering what I term "the neurologic causes" of a constricted visual field, there is a wide differential diagnosis including: bilateral occipital infarct with macular sparing, CRAO with cilioretinal sparing, post papilledema atrophy, a very miotic pupil, malingering/hysteria, and Cancer Associated Retinopathy (CAR). To remember this differential diagnosis for constricted visual fields:

Think of an old **SPARE tire** which looks like a ring scotoma and forms a constricted visual field. The tire is **ATROPHIC** and balding, and has many lumps and bumps in it that look like **DRUSEN**. A boy named **BOI** who is a **POOR PUPIL** because he is skipping school is playing with the tire and **PRETENDS** he is a **CAR**. (See Illustration)

SPARE tire	- bilateral occipital infarct with macular **SPARING**
	- CRAO with cilioretinal **SPARING**
ATROPHIC	- post papilledema **ATROPHY**
DRUSEN	- optic disc **DRUSEN**
BOI	- **B**ilateral **O**ccipital **I**nfarct (if you forget above)
POOR PUPIL	- very miotic **PUPIL**
PRETENDS	- malingering/hysteria
CAR	- **C**ancer **A**ssociated **R**etinopathy

Note: For the purpose of this discussion and the mnemonic, I have taken the liberty of classifying the other causes of constricted visual fields (glaucoma, retinitis pigmentosa, and high hyperopia) as "the non-neurologic causes" of constricted visual fields.

Note: The differential diagnosis for nyctalopia has much in common with the differential diagnosis for constricted visual fields as discussed above.

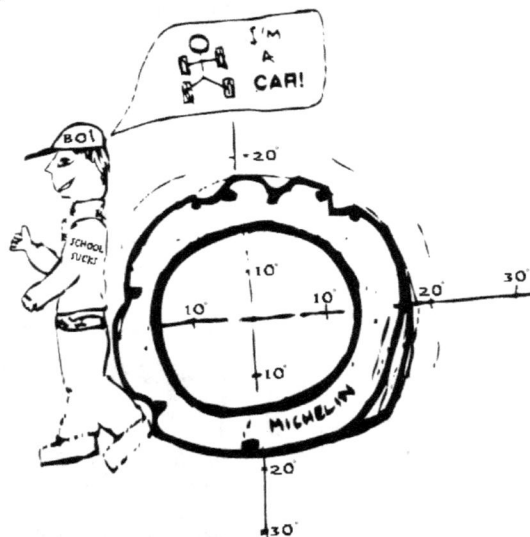

ELEVATED ESR IN FEMALES In females the Westergren erythrocyte sedimentation rate is considered elevated if it is greater than the quantity (patient's age +10) divided by 2. To remember the formula for elevated ESR in females:

Think – **TAB**, the abbreviation for temporal artery biopsy.

That is **TAB** - **T**en **A**dded **B**efore dividing by 2.

PRE-OP CONSIDERATIONS IN TEMPORAL ARTERY BIOPSY Before proceeding to the OR for temporal artery biopsy, go through this checklist. Ensure that the patient is not on anticoagulants other than aspirin. Compress the superficial temporal arteries to ensure that they are not critical collaterals, which on ligation, may cause stroke. The patient should be told of the possible need for bilateral biopsies, the risks of hemorrhage & delayed hemorrhage, and the risk of facial nerve palsy. Local anesthetic without epinephrine, cautery, and a doppler ultrasound should be available. The ESR, Fluorescein Angiography (FA), and oculoplethysmography (if available) results may be helpful in determining the likelihood of temporal arteritis, and need for biopsy.

Remember - The **ABCDEF checklist** before temporal artery biopsy.

Anticoagulation should be stopped
Bilateral **B**iopsy discussed &
Bleeding risk discussed (even delayed hemorrhage)
Compress the artery to exclude **C**ritical **C**ollaterals
Cautery available
Doppler available
Epinephrine should be avoided, **E**SR available
Facial nerve palsy discussed, **F**A available

GABA VS. GLUTAMATE It is important to know the excitatory versus inhibitory transmitter in the brain (e.g. basal ganglia). The GABA receptors respond to the neurotransmitter γ-aminobutyric acid (GABA), the chief inhibitory neurotransmitter in the central nervous system.

Glutamate is the most prominent neurotransmitter in the body, being present in over 50% of nervous tissue. The primary glutamate receptor is specifically sensitive to N-Methyl-D-Aspartate (NMDA) which is *excitatory.*

Think – Glu**+**amate.

Every time you see the word "glutamate" change the "t" to a "**+**" sign. The "**+**" in Glu**+**amate indicates that it is the excitatory neurotransmitter.

GABA sounds like "**gabbing.**"

When you're gabbing (always talking) it <u>inhibits</u> you from getting work done, just as **GABA** is the inhibitory neurotransmitter of the CNS.

ESSENTIAL BLEPHAROSPASM

Blepharospasm and hemifacial spasm may at first appear similar clinically. However, blepharospasm may be bilateral and can be associated with basal ganglia dysfunction. Botulinum toxin injection (trade name Botox) is the treatment for both blepharospasm and hemifacial spasm. However, blepharospasm often requires larger doses of Botox with a briefer average effect (3 months for blepharospasm vs. 4 months for hemifacial spasm). If you are having trouble distinguishing blepharospasm from hemifacial spasm:

Remember – The **6 B's** of **B**lepharospasm.

Blepharospasm
Bilateral
Basal ganglia dysfunction
Botox is the treatment:
Requiring **B**igger doses
With a **B**riefer average effect (than for hemifacial spasm)

Information source: American Academy of Ophthalmology Basic and Clinical Science Course: Neuro-Ophthalmology. American Academy of Ophthalmology, 2008, p. 287.

GENICULATE BODIES

The lateral geniculate body is for visual system and connects with the superior colliculus. The medial geniculate body is for the auditory system and is a relay between the inferior colliculus and the auditory cortex.

MAI – LES

Medial = **A**uditory to **I**nferior colliculus
Lateral = **E**ye to the **S**uperior colliculus

Mnemonic by permission of and modified from Robert O'Connor, MD, www.MedicalMnemonics.com. Copyright www.MedicalMnemonics.com.

CEREBRAL ANEURYSMS

Cerebral aneurysms are potentially deadly dilations of the cerebral circulation, usually involving the vessels of the Circle of Willis.

The rule of 10's for cerebral aneurysms.

10% are multiple.

10% are infratentorial.

> 10 mm aneurysms are most likely to rupture.

10% of CT scans may not show intraparenchymal or subarachnoid blood with cerebral aneurysm rupture, and in these cases, lumbar puncture should be performed.

The initial **10 day** period (4-**10** days) is the peak onset for vasospasm after a cerebral aneurysm ruptures.

About **10%** mortality (actually 15%) at **10** years for untreated aneurysms.

Information source: American Academy of Ophthalmology Basic and Clinical Science Course: Neuro-Ophthalmology. American Academy of Ophthalmology, 2008, p. 350.

EHLERS-DANLOS SYNDROME TYPE 4

Ehlers-Danlos syndrome type 4 (ED 4) is the most lethal of the 9 types. Patients with ED 4 may present to the ophthalmologist with spontaneous carotid-cavernous fistula. There is a 25% mortality rate directly attributable to diagnostic or therapeutic interventions for spontaneous carotid cavernous fistula. Although carotid cavernous fistula is vision threatening, it is rarely a life-threatening condition. Consequently the decision to proceed with diagnostic and therapeutic intervention versus observation must be weighed carefully.

Be **FORE**-warned about Ehlers-Danlos type **FOUR**.
One **FOUR**-th (25%) of the patients might die from the complications of femoral angiography.

Information source: Schievink WI et al. J Neurosurg, 1991; 74: pp. 991-998.

THE PULFRICH STEREO-ILLUSION

The Pulfrich phenomenon results from differential conduction rates between normal and damaged optic nerves. If a pendulum is swung in a straight line in front of a patient with a unilateral or asymmetric optic neuropathy, the pendulum may appear to rotate. The pendulum appears to rotate counterclockwise if the right optic nerve is more damaged and clockwise if the left optic nerve is more damaged.

For the Pulf**R**ich stereo-illusion,
Right optic neuropathy causes counte**R**clockwise rotation.

CRANIAL NERVE BASICS

Those of you that teach and interact with medical students may find the following helpful.

1) Nerves of Ocular Motility

The lateral rectus (L.R.) is innervated by the abducens nerve (CN VI). The superior oblique (S.O.) is innervated by the trochlear nerve (CN IV). The other muscles of extraocular motility are innervated by the oculomotor nerve (CN III). Remember the nerves of ocular motility with this little rhyme:

L.R. Six, S.O. Four,
All the rest are oculomotor.

Mnemonic origin unknown.

2) Upper vs. Lower Motor Neuron Lesions of Cranial Nerve 7

The upper face has bilateral innervation from the two cerebral hemispheres. Therefore in upper motor neuron facial nerve lesions, the forehead is spared. In lower motor neuron lesions both the upper and the lower face are paretic.

In **UPPER** motor neuron CN 7 lesions, the **UPPER** face stays **UP**. But, in **LOWER** motor neuron CN 7 lesions, the entire hemiface is **LOWER**.

3) The Trigeminal Nerve

The fifth cranial nerve is the trigeminal nerve.

The "**V**" in "CN V" looks like the bottom part of a GEM - The tri**GEM**inal nerve.

Mnemonic origin unknown.

4) Pupil-Involving Third Nerve Palsy

In any case of pupil-involving third nerve palsy, cerebral aneurysm of the posterior communicating artery must be excluded. Remember the association with this little rhyme:

Aneurysms of the p. comm.
Are third nerve bombs!

That is to say, aneurysms of the posterior communicating artery (**p. comm.**) are like little time bombs just waiting to blow the pupil!

NEURO-OPHTHALMIC EMERGENCIES

If you can't explain the patient's visual problem, don't forget the most important neuro-ophthalmic emergencies/urgencies.

DA₃M₃P₃ERS

Don't let a missed neuro-ophthalmic diagnosis put the **DA$_3$M$_3$P$_3$ERS** on your day!

Note: In this mnemonic there are three "**A**'s," three "**M**'s," and three "**P**'s" in **DA$_3$M$_3$P$_3$ERS**.

Dissection (carotid artery dissection with Horner's syndrome)

Arteritis (giant cell arteritis)
Aneurysm (pupil involving third nerve palsy usually)
Asystole (Kearns-Sayre Syndrome with heart block)

Myasthenia
Malignancy
Mucormycosis

Papilledema
Poisons (e.g. botulism)
Pituitary (Pituitary Apoplexy and
Panhypopituitarism in pediatric patients with de Morsier's syndrome)

Embolus of **R**etina and
Stroke

NOTES & YOUR OWN MNEMONICS:

NOTES & YOUR OWN MNEMONICS:

VII. GLAUCOMA

GOLDMANN EQUATION

The Goldmann equation calculates intraocular pressure using aqueous humor formation, outflow facility, and episcleral venous pressure as the variables. The equation is as follows:

$$P_o = (F/C) + P_e$$

P_o is intraocular pressure
F is rate of aqueous formation
C is outflow facility
P_e is episcleral venous pressure

To remember the Goldmann equation, keep this little rhyme in mind:

F. over C. plus P.E.
Is the Goldmann equation for I.O.P.

FACILITY OF OUTFLOW

The normal value for facility of outflow is **0.28** mcl/min/mm Hg. To remember normal outflow facility:

Think of an "**8 ball** hyphema" that is composed of innumerable **#8 pool balls** floating in the aqueous humor. A pool cue with **2** fingers attached to the end is desperately trying to push all the **8 balls** out of the anterior chamber in an attempt to re-establish normal outflow facility of **0.28** mcl/min/mm Hg. (See Illustration)

The **2** fingers trying to push out the **#8 balls** of the **8 ball** hyphema remind you of normal outflow facility of **0.28** mcl/min/mm Hg.

AQUEOUS VOLUME: ANTERIOR VERSUS POSTERIOR CHAMBER

The anterior chamber contains, on average, 250 microliters of aqueous humor while the posterior chamber contains only 60 microliters.

The normal **AC** holds approximately **a quarter of a cc**.
While the normal **PC** holds about **a quarter of a quarter**.

That is to say, the normal anterior chamber holds 250 microliters (a quarter of a cc).
The normal posterior chamber holds 62.5 (approximately 60) microliters (about a quarter of quarter of a cc).

Mnemonic by permission of and modified from Graham Trope, MBBS, PhD, FRCS(C).
Information Source: James Collins, Ophthalmic Desk Reference. Raven Press, 1991, p. 597.

JUVENILE GLAUCOMA: GENETICS

The TIGR gene located on the long arm of chromosome 1 (chromosome 1q) is one of the genes responsible for juvenile onset open angle glaucoma.

Imagine Tony the **TIGER** walking around with **ONE** pool **CUE** poking children in the eye and giving them glaucoma.

This reminds you that the TIGR gene (Tony the **TIGER**) responsible for juvenile open angle glaucoma is located on chromosome chromosome 1q (**ONE CUE**).

PEDIATRIC GLAUCOMA: EPIDEMIOLOGY

The majority of pediatric glaucomas are isolated congenital glaucoma. Slightly more patients are boys, and bilateral involvement occurs in the majority of cases.

Remember – The approximations of **6** and **60** for pediatric glaucoma.

60% of pediatric glaucomas are isolated congenital glaucoma.
60% of pediatric glaucoma is diagnosed by **6** months of age.
About **60**% (actually **65**%) of patients are **6**oys.
6ilateral involvement occurs about **60**% of the time (actually 70%).

Note: the figures in parentheses are actual figures from the BCSC text.

Information source: American Academy of Ophthalmology Basic and Clinical Science Course: Glaucoma. American Academy of Ophthalmology, 2008, p. 155.

NEURAL RIM THICKNESS IN NORMAL EYES

The optic disc is slightly oval with its vertical dimension slightly greater than its horizontal. The optic cup, on the other hand, has a horizontal oval shape in the majority of normal eyes and is located slightly off-center within the optic disc. Therefore, a vertical cup-to-disc ratio greater than the horizontal cup-to-disc ratio should be viewed with suspicion. This combination of the oval, off-center optic cup sitting within the oval optic disc causes a characteristic pattern of thickness in the surrounding neural rim in normal eyes. The inferior neural rim is the broadest (thickest) of the four quadrants, followed by the superior, nasal, and temporal rims.

If the **I**nferior neural rim **ISN'T** the broadest, then something **ISN'T** right.

The word "**ISN'T**" reminds you of the order of normal neural rim thickness from the broadest to thinnest in normal eyes: (broadest/thickest) **I**nferior, **S**uperior, **N**asal, then **T**empora rim (thinnest). If the neural rim does not follow the **ISN'T** rule, you should suspect glaucoma.

Mnemonic by permission of and modified from Dai Tran, MD and Elliot Werner, MD.

Information source: 1. MB Shields, Textbook of Glaucoma, 5th edition. Lippincott, Williams & Wilkins, 2005, p. 87.
2. J.B. Jonas et al, IOVS, 1988: 29; p. 1151.

ANGLE ANATOMY

From anterior to posterior the angle structures seen on gonioscopy are Schwalbe's line, the trabecular meshwork, Schlemm's canal (in some cases), scleral spur, and the ciliary body. An ophthalmologist won't pull out his hair trying to identify angle structures if he remembers:

Swell **T**rabecular **M**eshworks **C**an **S**pare **C**ilia.

Swell	**S**chwalbe's line
Trabecular **M**eshworks	**T**rabecular **M**eshwork
Can	**C**anal of Schlemm
Spare	**S**cleral **S**pur
Cilia	**C**iliary Body

TYPES OF GONIO LENSES

There are several gonio lenses that can be used to view the structures of the angle. To remember all the lenses:

Remember - The **Go**nio **KoPZS**.

Think of gonioscopy policemen or cops (**KoPZS**) whose job it is to ensure that you use at least one of the gonio lenses on every glaucoma patient.

Goldmann 3 mirror
Koeppe
Posner
Zeiss
Sussman (like a Goldmann 3 mirror with a narrow Zeiss base for indentation)

THINGS TO NOTE IN GLAUCOMA

When assessing a patient's risk for glaucoma keep the following risk factors and exam findings in mind:

The ABC's of glaucoma (all the way through "K").

Age (older age)
Black race
Cup-disc ratio & **C**entral **C**orneal thickness
Diabetes & **D**rance hemorrhages
Exfoliation
Field loss
Genetics (hereditary predisposition or family history)
Hyperopia
Intraocular pressure
Junctional scotoma (not to be confused with glaucoma)
Krukenberg spindle

Mnemonic by permission of and modified from Ravi Krishnan, MD.

PRIMARY OPEN ANGLE GLAUCOMA: RISK FACTORS

Several studies have elucidated many risk factors for the development of primary open angle glaucoma POAG. They include: family history, hypertension, a thin central corneal thickness (CCT), advanced age, African American race, myopia, diabetes, and female sex (although not every study agrees). To remember these risk factors:

Think - Your risk of POAG depends on the **Family CARDS** you're dealt.

Family history
Cardiovascular problems (specifically hypertension) &
Central **C**orneal thickness (a thin CCT)
Advanced **A**ge (higher risk with more advanced age)
Race (African American race) &
Refraction (POAG more common in myopes)
Diabetes
Sex (slightly higher risk in females)

Mnemonic by permission of and modified from Jeff Martow, MD, FRCS(C).

CENTRAL CORNEAL THICKNESS AND GLAUCOMA

Central corneal thickness is inversely related to the development of chronic open angle glaucoma and is a powerful predictor for the development of chronic open angle glaucoma. There is ample evidence that thinner corneas have falsely lower eye pressure readings.

Having a thin cornea is like skating on thinner ice.

OR

Eyes with thinner corneas may bust (i.e. have problems).

OCULAR HYPERTENSION: TREATMENT VERSUS OBSERVATION

At present, not all patients with ocular hypertension are treated. The risk of progression to glaucoma in ocular hypertensives is approximately 1% per year. Although this risk may seem relatively low, there are certain situations where you should cover the ocular hypertensive patient with IOP lowering glaucoma medications.

There are certain ocular hypertensives you should **COVER**.

Can't follow, **C**entral **C**orneal thickness (a thin CCT)
One-eyed
Vein occlusion (or IOP ≥ 30 due to increased risk of **V**ein occlusion)
Edema of cornea
Risk factors present (the **Family CARDS** mentioned above)

Mnemonic by permission of and modified from Jeff Martow, MD, FRCS(C).

UNILATERAL GLAUCOMA

When a patient presents with unilateral glaucoma or severely asymmetric disc cupping or visual field loss, be sure that your differential diagnosis extends well beyond POAG.

If a patient has unilateral glaucoma, they require a **UNeLATERAL FiX**.

Uveitis and iritis (including Posner-Schlossman syndrome)
Neoplasm (specifically ciliary body melanoma)
Lens **A**bnormalities (lens dislocation or phacolytic/phacomorphic/phacoanaphylactic glaucoma)
Trauma (hyphema, angle deformity, foreign body, or chemical burn)
Early POAG (asymmetric)
Rubeosis iridis
After **L**ong steroid drop use in one eye

Fistula (carotid cavernous/dural fistula)
Xfoliation (the pseudoexfoliation syndrome)

PLATEAU IRIS

With plateau iris, the ciliary processes are either rotated anteriorly or anteriorly displaced. They push anteriorly against the peripheral iris tissue tenting it up into and obstructing the angle and its associated drainage structures. To remember plateau iris configuration, remember:

The word **PLATEAU** describes the anatomy perfectly!

In **PLATEAU** iris, the **P**rocesses **L**ie **A**nterior **T**o **E**xpected **A**natomic **U**sualness.

NORMAL TENSION GLAUCOMA: DIFFERENTIAL DIAGNOSIS

Patients may present with what appears to be glaucomatous field loss or cupping, but without ever having an elevated pressure measured in the office. Before simply diagnosing these patients with normal tension glaucoma, take a detailed history and perform a thorough exam with additional testing, if necessary, to rule out other causes of apparent disc cupping and/or field loss in the presence of normal office-measured pressures. To remember the differential diagnosis for normal tension glaucoma:

Remember your **ConPreSShION CloVar** for normal tension glaucoma! (See Illustration)

Think about carrying your lucky **ConPreSShION CloVar** ("compression clover") around in your head so you will be lucky and never mistake a case of optic nerve compression for normal tension glaucoma.

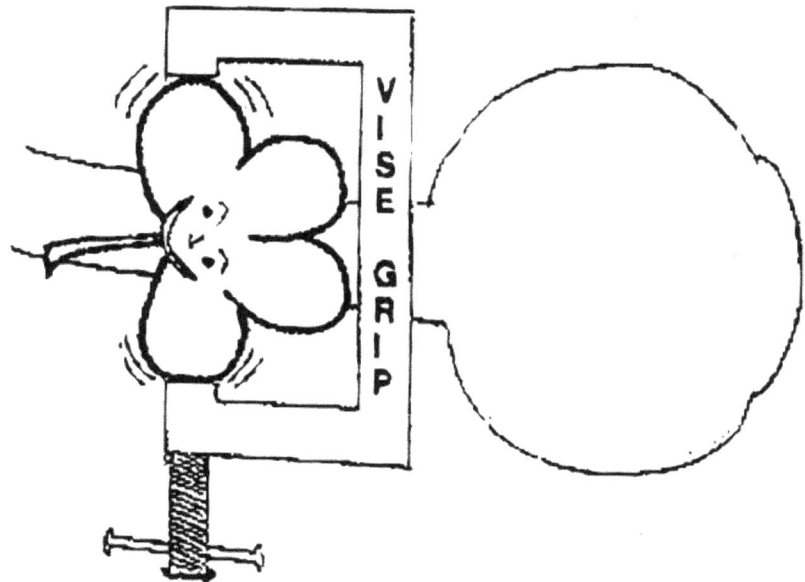

ConPreSShION (optic nerve compression)

Congenital lesions (coloboma, optic nerve pits, and optic disc drusen (although not really "congenital"))
Previous glaucoma now controlled (as occurs in burnt out glaucoma, burnt out pigment dispersion glaucoma, and uveitic glaucoma)
Steroids (prior steroid use)
Shock (myocardial infarction, anemia, and hemorrhage)
Iritis including Posner-Schlossman
Optic **N**europathy (Traumatic, Inflammatory, Arteritic, or NAION)

Closure of the angle, intermittent
Variable pressure (diurnal variation)

NEOVASCULAR GLAUCOMA & RUBEOSIS IRIDIS

The three most common causes of rubeosis iridis are diabetes mellitus, central retinal vein occlusion, and carotid disease. Carotid disease is sometimes forgotten, but is easier to recall with this mnemonic for the differential of rubeosis iridis:

Think - **CAROTID DS.** causes rubeosis iridis.

That is **CAROTID DS.** ("carotid disease") causes rubeosis iridis for the differential of rubeosis.

Carotid stenosis & fistula, **C**RVO, **C**oat's disease
Arteritis (including Giant cell & Takayasu's), **A**rterial occlusion
Radiation, **R**OP (retinopathy of prematurity)
Occlusion of the vein or artery (if you forgot the above)
Trauma, **T**umor (including retinoblastoma & melanoma)
Inflammation (uveitis), **I**diopathic (Eales)
Detached retina

Diabetes mellitus (with proliferative retinopathy or post-vitrectomy)
Sickle cell hemoglobinopathy

THE IRIDOCORNEAL ENDOTHELIAL SYNDROMES

The iridocorneal endothelial syndromes are often referred to as the ICE syndromes. The name "ICE syndromes" is a quick mnemonic to remember the three diseases that make up the ICE syndromes:

The **ICE** syndromes are:
Iris nevus syndrome, **C**handler's disease, and **E**ssential iris atrophy

---

MANUAL VERSUS AUTOMATED PERIMETERS

Perimeters can be either controlled by hand (manual) or controlled by a computer (automated). The Humphrey perimeter is an example of an automated perimeter, while the Goldmann perimeter is a manual perimeter. If your staff can't remember whether a Goldmann field is a manual or automated field just remind them:

A Gold**MAN**n perimeter is a **MAN**ual perimeter.

ARCUATE VISUAL FIELD DEFECTS

The Bjerrum scotoma is an arcuate scotoma, corresponding with the arcuate distribution of the retinal nerve fibers. It extends from the blind spot all the way to the median raphe (either above or below). A Seidel scotoma is an early arcuate defect that extends in a curved course from blind spot toward the median raphe, but it does not reach all the way to the raphe. Instead, the Seidel scotoma tapers to a point before reaching it.

The **B**jerrum defect is **B**ig.
The **S**eidel **S**cotoma is **S**mall.

This reminds you that the **B**jerrum defect is **B**ig enough to extend all the way from the blind spot to the median raphe, while the **S**eidel **S**cotoma is so **S**mall it cannot reach the median raphe and simply tapers off to a point.

STIMULUS SIZE IN FORMAL PERIMETRY

The AREA of the standard stimulus sizes are as follows:

0	=	1/16 (0.0625) mm^2
I	=	1/4 (0.25) mm^2
II	=	1 mm^2
III	=	4 mm^2
IV	=	16 mm^2
V	=	64 mm^2

mnemonic

The size of the spot (area) is **4 to the power of (n-2)** where **n** is the Roman numeral designation of the spot.

For example: Stimulus size I = $4^{(1-2)} = 4^{-1}$ = 1/4 or 0.25 mm^2
Stimulus size IV = $4^{(4-2)} = 4^2$ = 16 mm^2

ADRENERGIC RECEPTORS & PUPIL DILATION

The alpha 1 receptors mediate pupil dilation.

mnemonic

Picture dilating the patient's pupils by using a bottle of **A-1 steak sauce** in the eyes. (See Illustration)

Thus **Alpha 1** receptors (the **A-1 steak sauce**) mediate pupil dilation.

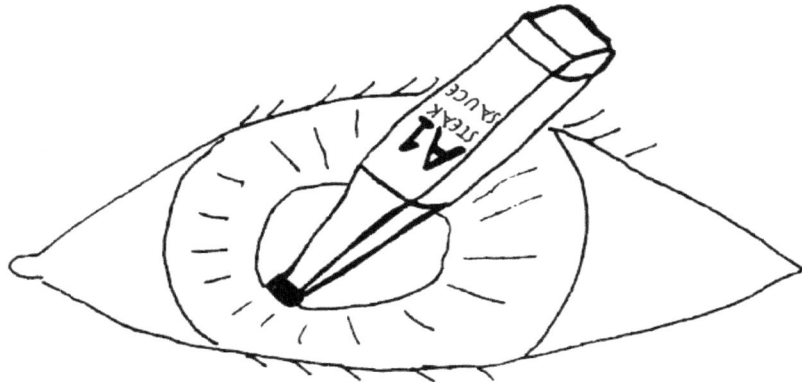

THE ALPHA AGONIST AND ANTAGONIST

Thymoxamine (moxisylyte) is an alpha antagonist that is mainly of historic interest. Thymoxamine is a miotic and was shown to lower intraocular pressure. Apraclonidine, on the other hand, is an alpha 2 agonist. Although it may seem counterintuitive, both the alpha antagonist thymoxamine and the alpha 2 agonist apraclonidine lower intraocular pressure! To remember the mechanism of action of each drug, remember this little rhyme:

mnemonic

**Thymox alpha blocks,
But Apraclon turns alpha 2 on.**

AQUEOUS SUPRRESSANTS

The majority of topical anti-glaucoma medications decrease intraocular pressure by decreasing the production of aqueous humor. These include the topical alpha 2 adrenergic agonists, beta blockers, and carbonic anhydrase inhibitors. In addition, the alpha 2 adrenergic agonists are believed to have an additional mechanism of action. While both alpha 2 adrenergic agonists decrease aqueous production, apraclonidine (trade name Iopidine) decreases episcleral venous pressure while brimonidine (trade name Alphagan) increases uveoscleral outflow. To remember the mechanism of action of these aqueous suppressants:

The aqueous suppressants are **A,B,C**

That's **A**lpha 2 agonists, **B**eta blockers and **C**arbonic anhycrase inhibitors – the aqueous suppressants.

OR

**Alpha (+), Beta (-), C-A-I,
All decrease production of aquei**.

Alpha (+) – alpha 2 adrenergices
Beta (-) – beta blocker
C-A-I – Carbonic Anyhdrase Inhibitors
aquei – aqueous humor

AND

The alpha **TWO** adrenergics have **TWO mechanisms of action**.

The first mechanism of action is aqueous suppression.
The second mechanism of action is different fro each agent:
Apraclonidine (trade name Iopidine) also decreases episcleral venous pressure.
Brimonidine (Alphagan) also increases uveoscleral outflow.

Information source: American Academy of Ophthalmology Basic and Clinical Science Course: Glaucoma. American Academy of Ophthalmology, 2008, pp. 170-174.

TRADE NAME FOR BRIMONIDINE

Brimonidine (trade name Alphagan by Allergan) is an alpha 2 adrenergic agonist, not to be confused with bimatoprost (trade name Lumigan by Allergan) a prostaglandin analogue.

To play your "**Alpha Game**" keep the "**Brim On**" your best hat!

"**Alpha Game**" sounds like "**Alphagan**."
"**Brim On**" sounds like "**brimonidine**."
This reminds you that brimonidine "**Brim On**" is the generic name for the drop trade-named Alphagan "**Alpha Game**."

AND

Alphagan is an **Alpha** 2 adrenergic agonist.
It is NOT the same class of drug as bimatoprost (trade name Lumigan) a prostaglandin analogue by Allergan.

PROSTAGLANDIN ANALOGUES: TRADE & GENERIC NAMES

1) The prostaglandin analogues are considered by many to be the first-line in topical glaucoma therapy. The trade and generic names of each are as follows: Xalatan (latanoprost), Lumigan (bimatoprost), Travatan (travoprost), and Rescula (unoprostone). Remember the prostaglandin analogue trade names with this little rhyme:

Prostaglandins to the rescue,
Trade names "XaLuTrRescu!"

Xa = **Xa**latan (Pfizer)
Lu = **Lu**migan (Allergan)
Tr = **Tr**avatan (Alcon)
Rescu = **Rescu**la (Novartis)

2) Once you know the trade names of each of the prostaglandin analogues, it is easy to find their generic names (usually hidden within the trade name). Let us look at each starting with the trade name (in plain print) and then the generic name (in italic print).

Xa**LATAN** = *LATANoprost*
The words Xa**LATAN** and *LATANoprost* have many similar letters.

TRAVatan = *TRAVoprost*
The words **TRAV**atan and *TRAVoprost* also have many similar letters.

Lumigan = *Bimatoprost*
Think – "**Lumi-Bim**."
"**Lumi-Bim**" sounds like "**loony bin**" reminding you that **Lumi**gan = *Bimatoprost*

Rescula = *Unoprostone*
Remember – **RESCU**e numero **Uno**.
RESCUe numero **Uno** reminds you that **Rescu**la = *Unoprostone*.

ARGON LASER TRABECULOPLASTY

Argon laser trabeculoplasty should be performed with a 50 micron spot size, using 50 applications over 180 degrees of the angle. 50% of patients with an initial response maintain a lower pressure 5 years post laser. Hence:

The Rule of 50's for ALT.

50 micron spot size
50% of the angle is lasered at initial sitting (180 degrees)
50 applications
50% of patients with an initial response maintain a lower pressure 5 years post laser

Information source: American Academy of Ophthalmology Basic and Clinical Science Course: Glaucoma. American Academy of Ophthalmology, 2008, p. 189.

THE CARDIOSELECTIVE BETA BLOCKER

Betaxolol (trade name Betoptic) is the only available topical beta blocker that is cardioselective and should theoretically result in less bronchospasm. This would make betaxolol the beta blocker of choice in patients with potential asthma or COPD. Please keep in mind however, that all beta blockers should be avoided or used with extreme caution in patients with known asthma or COPD, whether cardioselective or not.

mnemonic — Be**top**tic is **tops** because it doesn't **tax** (be**tax**olol) breathing.

ADVERSE EFFECTS OF PROSTAGLANDINS

Latanoprost (trade name Xalatan) is a prostaglandin analogue used in the treatment of glaucoma. Known adverse effects of prostaglandin analogues include increased iris and skin pigmentation, macular edema, and lengthening of the cilia.

mnemonic — Remember - **LaTanOProst** for adverse effects of the prostaglandins.

La - **La**shes Lengthen
Tan – "**Tan**ning" or darkening of the iris and skin
O – **O**edema of the macula (if you don't' mind using a little British English)
Prost - **Prost**aglandin analogues

METABOLISM OF ORAL OSMOTIC AGENTS

Oral osmotic agents can be used to lower intraocular pressure acutely in patients with angle closure or in other emergency situations where routine medications have proven inadequate. The two oral osmotic agents are isosorbide and glycerol. Glycerol should be avoided in diabetics since it is metabolized to glucose and can cause hyperglycemia.

mnemonic — **Iso**sorbide keeps the patient **Iso**glycemic,
While **Gl**ycerol is metabolized to **Gl**ucose.

RIMEXOLONE AND INTRAOCULAR PRESSURE

Rimexolone 1% (trade name Vexol) is a steroid ophthalmic preparation that has been shown to induce less intraocular pressure rise than 0.1% dexamethasone sodium phosphate or 1.0% prednisolone acetate.

mnemonic — **RIM-EX**olone keeps the optic **RIM-EX**cellent.

CONJUNCTIVAL CLOSURE FOR TRABECULECTOMIES

Taper point needles, which are round with no cutting surfaces, are used for conjunctival closure in trabeculectomies to reduce the incidence of leaking blebs by minimizing cutting of the conjunctiva by the needle. For trabeculectomies remember:

mnemonic — **TAPE**r point needle helps to **TAPE** the conjunctiva closed.

TYPES OF SETONS

Seton implants, commonly referred to as tube shunts, are glaucoma drainage implants than consist of a tube which is inserted into the anterior chamber and a fluid reservoir which helps disperse the fluid beneath the conjunctiva. Certain models contain a valve, others do not. The different types of setons include the Joseph, Ahmed, Molteno, Baerveldt, Krupin, and Schocket. To remember these names:

Think of the Seton **JAM BoKS**.

Each time you sit down at a diner and see the little plastic boxes of **JAM**, imagine each one is a little Seton **JAM BoKS**.

Joseph
Ahmed
Molteno

Baerveldt
o
Krupin
Schocket

NOTES & YOUR OWN MNEMONICS:

NOTES & YOUR OWN MNEMONICS:

VIII. RETINA

THE DARK CURRENT

In the dark, rods have open sodium channels. A net influx of sodium ions (Na^+) results in a continuous current called the dark current, which maintains the rods in a depolarized state.

Dark Curr**EN**t

Depolarized rods due to **Ent**ry of **Na**$^+$

Information source: Robert Berne & Matthew Levy (eds), Physiology, 3rd edition. Mosby Year Book, 1993, p. 151.

GLYCOSYLATED HEMOGLOBIN

In recent years, the level of glycosylated hemoglobin (Hb A1c) has become a popular and valuable marker for evaluating blood sugar control in diabetics. Normal Hb A1c is 7% or less.

Remember - **7** is your lucky number,
And if you're lucky your **HbA1c** will be **7** or less.

CLINICALLY SIGNIFICANT MACULAR EDEMA

The three criteria for clinically significant macular edema (CSME) are macular edema within 500 microns of the foveola, hard exudates within 500 microns of the foveola with accompanying macular edema, or macular edema of at least one disc diameter in size (1500 microns) within one disc diameter of the foveola. Remember the criteria for CSME with this little rhyme:

ME, HE-ME, or ME X 3
In units of 500, from the fovee.

ME	**M**acular **E**dema within **500** microns of the foveola
HE-ME	**H**ard **E**xudates within **500** microns of the foveola with associated **M**acular **E**dema
ME X 3	**M**acular **E**dema of 1500 microns (one disc diameter) within 1500 microns of the foveola
In units of 500	**500** microns
from the fovee	From the foveola

Note: One disc diameter is 1500 microns.

THE FLECKED RETINA

Multiple yellow dots throughout the fundus can be categorized into a group of conditions described anatomically as "the flecked retina." The differential includes dominant drusen, retinitis punctata albescens, fundus albipunctata, Kandori's flecked retina, and Stargardt's disease (a.k.a. fundus flavimaculatus).

Unless you learn the differential for a flecked retina they'll call you, "**DR. FAK**e**S**!"

Dominant **D**rusen
Retinitis punctata albescens
Fundus **A**lbipunctata
Kandori's flecked retina
e
Stargardt's disease (a.k.a. fundus flavimaculatus)

CAUSES OF CYSTOID MACULAR EDEMA

Cystoid macular edema is associated with systemic diseases such as diabetes, local conditions such as pars planitis, retinitis pigmentosa, or venous obstruction, drugs such as epinephrine, E2 prostaglandin analogues, and nicotinic acid, and post-cataract surgery as in Irvine-Gass syndrome.

mnemonic — Remember – **DEPRIVEN'** for the causes of macular edema.

Think of macular edema **DEPRIVEN'** ("depriving") patients of their vision.

Diabetes
Epinephrine
Pars planitis
RP (retinitis pigmentosa)
Irvine-Gass syndrome
Venous obstruction
E2 prostaglandin analogues
Nicotinic acid

Mnemonic by permission of and modified from Gary Brown, MD.

SEVERE NON-PROLIFERATIVE DIABETIC RETINOPATHY

Patients with Severe non-proliferative diabetic retinopathy (Severe NPDR) are a high risk for progression to proliferative retinopathy. The Early Treatment of Diabetic Retinopathy Study (ETDRS) identified severe non-proliferative diabetic retinopathy as any fundus that met at least one of these three criteria: 4 quadrants of microaneurysms or hemorrhages, 2 quadrants of venous beading, or 1 quadrant of intraretinal microvascular abnormalities (IRMA). This 4:2:1 rule helps define those patients at greatest risk for progression to proliferative diabetic retinopathy. If you use this mnemonic every time you write the words "aneurysm," "venous beading," and "IRMA," you'll never forget the criteria for severe NPDR.

mnemonic — **4**neurysms or hemorrhages

✌ enous beading

1RMA

Use an italicized numeral "**4**" to make the letter "A" in the word aneurysm, reminding you of the criterion, 4 quadrants of microaneurysms or hemorrhages.

Use two fingers to make the "V for Victory sign" in the words venous beading, reminding you of the criterion, 2 quadrants of venous beading.

Finally, use the numeral "**1**" to make the letter "I" in the word IRMA, reminding you of the criterion, 1 quadrant of IRMA.

APPEARANCE OF X-LINKED RETINOSCHISIS

The characteristic appearance of juvenile (X-linked) retinoschisis is foveal schisis with radial linear folds running from the center of the fovea outward.

mnemonic — X-linked retinoschisis is a foveal schisis that looks like a whole bunch of **X**'s across the fovea.

SUBRETINAL NEOVASCULAR MEMBRANES: DIFFERENTIAL DIAGNOSIS

The most common cause of subretinal neovascular membranes (a.k.a. choroidal neovascular membranes) is age related macular degeneration. However, there are other common and less common causes of this condition. To remember the differential diagnosis for subretinal neovascular membranes:

Imagine that subretinal damage is caused by **SUB-MARINe HISTO**. (See Illustration)

SUB-MARINe HISTO is an aged (ARMD) and damaged sub that streaks (angioid streaks) underneath the retina (subretinal) and wrecks everything in its path. The submarine is out of control because it is run by a highly myopic idiot (high myopia & idiopathic). The name of the sub boldly flashes in neon (neoplasm) lights.

SUB-retinal neovascular membranes

Myopia (high myopia)
Age **R**elated macular degeneration
Idiopathic
Neoplasms

HISTOplasmosis
Heredodegenerative diseases (Bests, Stargardt's, & drusen of optic nerve)
Inflammation
Streaks (Angioid)
Trauma (choroidal rupture, laser, & cryotherapy)
Optic nerve disease (drusen, pits, etc.)

ANGIOID STREAKS

Angioid streaks are commonly found in pseudoxanthoma elasticum, Ehlers-Danlos syndrome, Paget's disease of bone, and sickle cell anemia; or they may be idiopathic – thus the popular mnemonic **PEPSI** for the differential of angioid streaks. There are also less common disease processes associated with angioid streaks, some of which can be worked into the mnemonic.

Think – **PEPSI** for the differential of angioid streaks.

Imagine a **PEPSI** bottle shattering Bruch's membrane and causing angioid streaks. (The less common, but related associations in parentheses)

Pseudoxanthoma elasticum
Ehlers-Danlos (and **E**lastosis Senilis (a.k.a. Senile Elastosis))
Paget's disease of bone (and Hypercalcinosis and Hyperphosphatemia)
Sickle cell anemia (and other blood disorders - Hemolytic anemia, Hemochromatosis, and Hereditary spherocytosis)
Idiopathic

Mnemonic origin unknown.

RHEGMATOGENOUS RETINAL PATHOLOGY: THE RULE OF 7's

Patients often inquire about how common their retinal disease is or how did this happen to them. In answering these questions, here are some useful approximations.

The Rule of 7's for rhegmatogenous retinal pathology.

Approximately:
7% of the population has a retinal break.

7 per 100,000 people per year develop a retinal detachment.

0.0**7**% of the population will develop a retinal detachment if they live to **7**0 years of age.

7% of the population has lattice retinal degeneration; but lattice degeneration is NOT the cause of the retinal break in **7**0% or more of patients with retinal detachment.

Information source: American Academy of Ophthalmology Basic and Clinical Science Course: Retina & Vitreous. American Academy of Ophthalmology, 2008, p. 291.

SHAFER'S SIGN

Shafer's sign refers to small clumps of pigmented cells (tobacco dust) in the vitreous of patients with rhegmatogenous retinal detachment.

"Shafer" sounds like "**Shaver**"

Think of shavings of tobacco dust.
There is a rhegmatogenous detachment because the shaver has nicked the retina!

SALT & PEPPER RETINOPATHY

Classically, salt & pepper retinopathy is associated with congenital rubella syndrome. Other diseases, and even carrier states of certain diseases, can also present with a salt & pepper appearance to the fundus. To remember the differential of salt & pepper retinopathy:

Remember - **MRS. L. CooKS** with **salt & pepper**. (See Illustration)

salt & pepper = salt & pepper retinopathy

Medications (**M**ellaril, Thorazine, & clofazimine)
Rubella, **R**etinitis Pigmentosa (in early RP disease & X-linked recessive carriers)
Syphilis
Leber's congenital amaurosis
Carriers **o**f many **C**onditi**o**ns (**Cho**roideremia & albinism), **C**ystin**o**sis, **C**MV
Kearns-**S**ayre syndrome

Information source: Kenneth Noble & William Freeman (eds), Practical Atlas of Retinal Disease and Therapy. Lippincott-Raven, 1998, p. 3.

BULL'S-EYE MACULOPATHY Bull's-eye maculopathy is typically associated with diseases that preferentially destroy the cone photoreceptors such as cone dystrophy. The greatest imitator of bull's-eye maculopathy is age related macular degeneration with its predisposition for the posterior pole. Other causes of bull's-eye lesions include chloroquine and other antimalarials, benign concentric annular dystrophy, Stargardt's disease, and Spielmeyer-Vogt-Batten-Mayou syndrome (a cerebro-retinal degeneration syndrome of disturbed lipid metabolism). To remember the differential for bull's-eye maculopathy:

Think of a dead bull's **CAR-CASS** with a **bull's-eye** target painted on its hide. (See Illustration)

bull's-eye – bull's-eye maculopathy

Cone dystrophy
Age **R**elated macular degeneration
Chloroquine & other **A**ntimalarials, **C**oncentric **A**nnular dystrophy (benign)
Stargardt's
Spielmeyer-Vogt-Batten-Mayou syndrome

STARGARDT'S DISEASE: INHERITANCE

Stargardt's disease is <u>typically</u> an autosomal recessive disease. Although, some autosomal dominant cases have been reported.

St**AR**g**AR**dt's is typically **A**utosomal **R**ecessive.

The "**AR**'s" in St**AR**g**AR**dt's remind you it's typically **A**utosomal **R**ecessive.

THE CLASSIFICATION OF PROLIFERATIVE VITREORETINOPATHY	Proliferative vitreoretinopathy (PVR) is classified according to grades A, B and C. Grade C PVR is characterized by full thickness retinal folds and is further described by contraction type and anterior or posterior involvement.

Remember – The **ABC-CAP** of PVR.

The grades **A**, **B**, & **C** of PVR are just like the grades you get in school. For school and PVR, it is better to have an A grade than a B or C. The **CAP** stands for **C**ontraction type, **A**nterior extent, and **P**osterior extent. |
| **STICKLER'S SYNDROME VERSUS WAGNER'S DISEASE** | Stickler's syndrome and Wagner's disease are distinct vitreoretinopathies caused by different chromosomal defects.

1) In Stickler's syndrome there is a predisposition for rhegmatogenous retinal detachment and associated systemic findings such as arthropathy and Pierre Robin sequence.

Remember - The retina un**Stick**s in **Stick**ler's syndrome.

The retina un**Stick**s reminds you that patients with **Stick**ler's get retinal detachment. The "**S**" in **Stick**ler's reminds you of the associated **S**ystemic findings of arthropathy and Pierre Robin sequence in **Stick**ler's syndrome.

2) In Wagner's disease there is a markedly decreased likelihood of rhegmatogenous retinal detachment (although peripheral tractional RD may be present). There are ERG abnormalities in Wagner's disease.

Dogs with **WAG**ner's disease **WAG** their tails because they are less likely to require retinal surgery (as opposed to Stickler's syndrome).

The letters "**G**," "**E**," & "**R**" in the word Wa**G**n**ER**'s reminds you of the abnormal **ERG** in Wa**G**n**ER**'s disease.

Although both Wa**G**n**ER**'s and Stickler's end with "**ER**", only Wa**G**n**ER**'s disease has the "**G**" to make an abnormal **ERG** in Wa**G**n**ER**'s disease. |
| **CHOLESTEROLOSIS VERSUS ASTEROID HYALOSIS** | On exam, the opacities in cholesterolosis (a.k.a. synchisis scintillans) are crystals that often settle inferiorly since there is frequently an associated posterior vitreous detachment. However, in asteroid hyalosis, the calcific opacities are evenly distributed throughout the vitreous cavity.

Cholesterolosis **C**ollects **C**audally.

To remember that *synchisis scintillans* refers to cholesterolosis, think of a *scintillating* Hollenhorst embolus in a retinal vessel. |

TRAUMATIC ENDOPHTHALMITIS

Bacillus cereus accounts for almost 25% of cases of post-traumatic endophthalmitis.

Remember - Trauma causes **serious** endophthalmitis.

The word **"serious"** sounds like **"cereus,"** reminding you that *Bacillus cereus* is frequently associated with post-traumatic endophthalmitis (approximately 25% of the time).

American Academy of Ophthalmology Basic & Clinical Science Course: Retina & Vitreous. American Academy of Ophthalmology, 1996, p. 221.

CAUSES OF PERIPHERAL NEOVASCULARIZATION

There are a variety of causes of peripheral neovascularization, including the usual suspects like diabetes mellitus and BRVO. Then there are more obscure causes such as familial exudative vitreoretinopathy (FEVR) and Eales disease. Remember the full differential with this mnemonic.

In peripheral neovascularization there are
Usually **Sick PEROPHFEREALS** (as in "sick peripheral vessels").

Usual causes of neovascularization (DM, BRVO)
Sickle cell anemia (especially HbSC disease)
Pars planitis
Emboli
ROP (Retinopathy Of Prematurity)
Hyperviscosity syndrome
Familial **E**xudative vitreo**R**etinopathy (FEVR)
EALe**S** disease
Autoimmune diseases like **L**upus and **S**arcoid

FLUORESCEIN ANGIOGRAPHY OF CHOROIDAL FOLDS

Hyperfluorescence is seen at the crests of choroidal folds since the RPE cells there are splayed apart. This allows increased transmission of fluorescence at the crests. Hypofluorescence is seen in the troughs of choroidal folds, since the RPE cells there are more compressed. Thus the fluorescence is blocked at the troughs.

Hyper means **Higher**, and there's **Hyper**fluorescence at the **Crest**.
Hypo means **Lower**, and there's **Hypo**fluorescence at the **Trough**.
(See Illustration)

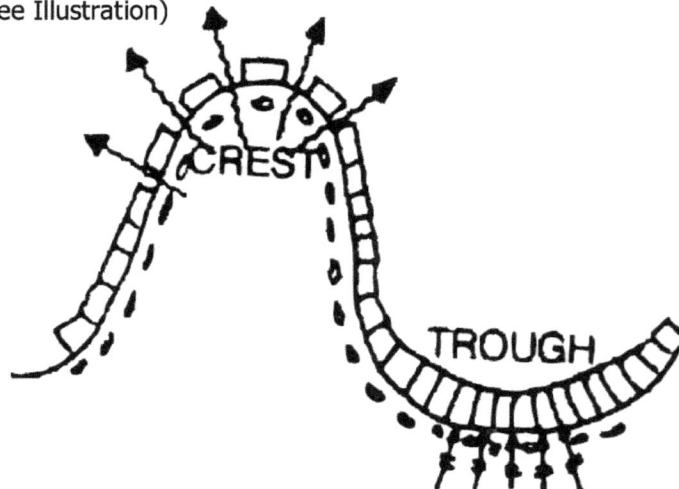

ACQUIRED RETINOSCHISIS VERSUS RETINAL DETACHMENT

On fundus exam, it may be difficult to differentiate a schisis cavity from a retinal detachment. If one looks closely, there are some clues on clinical exam that will indicate one has indeed found a schisis cavity.

HyperGun Skis are Absolutely Smooth and BITe No Dust or Lines.

First of all, the word "**schisis**" sounds like "**skis.**" The dome of retinoschisis is smooth and shaped like a ski hill. Imagine the hill covered by snowflakes made of little **Gunn's dots**. What brand of skis is best for this hill covered with snowflake Gunn's dots? **HyperGun** skis of course! This reminds you of "**Hyperopia**" and "**Gunn's Dots.**" They glide so absolutely **Smooth** over the dome of the retinoschisis ski hill that they don't **BITe** into the snow (**B**ilateral **I**nfero**Te**mporal schisis cavities), kick up no snow **Dust** (no tobacco **Dust**), and leave no **Lines** (no demarcation **Lines**).

Hyper	**Hyper**opic (often)
Gun	**Gun**n's dots (the footplates of Müller)
Skis	**Schis**is
Absolutely	**Absolute** visual field defect
Smooth	**Smooth**-domed schisis cavity
BITe	**B**ilateral, **I**nfero**Te**mporal schisis cavities
No Dust or Lines	**No** tobacco **Dust**, **No** demarcation **lines**

NYCTALOPIA

There are numerous causes of night blindness, including retinitis pigmentosa, glaucoma, and systemic medications such as phenothiazines, chloroquine, and quinine. Cancer-associated retinopathy (CAR) may present with nyctalopia, and although CAR is rare, it should always be considered in the differential diagnosis. Remember the differential diagnosis for nyctalopia with this little rhyme:

GLAUCOM_{AZ}, Meds, CAR, RP,
Make it hard at night to see.

GLAUCOM_{AZ} (advanced glaucoma)
Meds (phenothiazines, chloroquine, & quinine)
CAR (cancer-associated retinopathy)
RP (retinitis pigmentosa)

Gyrate atrophy
LAser (panretinal photocoagulation)
Unreal (hysteria, malingering, or functional vision loss)
Congenital stationary night blindness, **C**horoideremia, & **C**AR (mentioned above)
Optic atrophy (peripheral filed restriction), **O**guchi's disease (a form of CSNB)
Myopia, **M**iotics, & **M**eds (mentioned above)
A vitamin deficiency
Zinc deficiency

Note: The letters "**A**" and "**Z**" are depressed because they cause nyctalopia in deficiency.

A-SCAN ULTRASOUND: CHOROIDAL HEMANGIOMA VERSUS MELANOMA

It may be difficult to distinguish a choroidal hemangioma from a choroidal melanoma on clinical exam. A-scan ultrasound can help differentiate the two because choroidal hemangiomas typically have high internal reflectivity while choroidal melanomas typically have medium to low internal reflectivity.

mnemonic

Hemangiomas have **H**igh internal reflectivity.
MeLanomas have **M**edium to **L**ow internal reflectivity.

LASER WAVELENGTHS TO AVOID IN THE MACULA

The disadvantages of argon blue-green laser include: uptake by macular xanthophyll, potential photochemical toxicity associated with blue wavelengths, and increased scatter and absorption by cataractous lenses.

mnemonic

Avoid **BLUE** in the macula or you'll be singing the **BLUES**.

In other words, you must be very cautious when using blue-green argon laser because the macular xanthophyll will readily pick it up resulting in macular burns.

HORSESHOE TEARS OF THE RETINA

1) In a flap tear (a.k.a. horseshoe tear) the vitreous attached to a strip of retina pulls that strip of retina anteriorly, towards the ora serrata. Therefore, the base of the flap (i.e. where the flap is still attached to the retina) is more anterior. To draw a horseshoe tear correctly:

mnemonic

Remember – The Horse always walks toward the optic nerve.

That is to say, on your drawing, the horse track formed by the horseshoe tear should be oriented as if the horse is walking toward the optic nerve (with the open end toward the ora serrata, and the round, closed end toward the optic nerve).

2) When performing laser retinopexy to a horseshoe tear, the most important part to laser is the anterior base of the flap. Inadequate anterior laser treatment is the most common cause of retinopexy failure in flap tears. Continuing vitreous traction will eventually break through laser photocoagulation at the base of the flap if it is insufficient.

mnemonic

Imagine that an **Ant** is trying to pull up the flap and extend the tear! To stop spread of the tear, be sure to thoroughly laser the **Ant** at the **Ant**erior aspect of the tear. (See Illustration)

| **RETINAL DEGENERATION WITH HEARING LOSS** | Many diseases have retinal abnormalities and hearing problems. When these patients with retinal degeneration and hearing loss present to the clinic: |

RUB your thinking cap and **USHER**
them to the **COCK**-eyed* **HARD OF** hearing **WARD**.

RUB = Rubella
USHER = Ushers syndrome
COCK-eyed = Cockaynes syndrome

HARD OF
Hallgren's syndrome, **H**urler's syndrome
Alport's syndrome, **A**lstrom's syndrome
Refsum's syndrome
Dysplasia spondyloepiphysia congenita

Osteopetrosis
Friedrich's ataxia, **F**lynn-Aird syndrome

WARD = Waardenburg syndrome

*Note: Although these patients may not have strabismus (i.e. be "cock-eyed") they do have abnormal retinas with hearing loss.

Mnemonic by permission of and modified from Mike Myles, MD, FRCS(C).

CHOROIDAL MELANOMA VERSUS CHOROIDAL NEVUS

There are several distinct features that will distinguish a choroidal melanoma from a benign choroidal nevus. Findings in choroidal melanoma (that are notably absent in choroidal nevus) include: breaks in Bruch's membrane, retinal detachment, and elevation greater than 3 mm on ultrasound; orange flecks on clinical exam; multiple sites of pinpoint hyperfluorescence on FA; and lesion growth over time.

Remember – **BRUCH'S** for characteristics of choroidal melanoma.

Break in Bruch's membrane in melanoma
Retinal detachment (serous) in melanoma
Up (elevation > 3 mm suggests melanoma)
Color (orange flecks suggest melanoma)
Hyperfluorescenct multifocal pinpoints on FA of melanoma
        (but a **H**alo on FA suggests choroidal nevus)
Size change (growth) suggests melanoma

Mnemonic by permission of and modified from Ravi Krishnan, MD.

RADIATION RETINOPATHY

Radiation retinopathy may result after radiation treatment for local or regional disease to the head, orbit or eye. It can result in macular edema and proliferative or nonproliferative retinopathy similar to that found in diabetes mellitus. The onset of radiation retinopathy is typically delayed (around 15 months post treatment) and seen in doses in the range of 30 gray (3000 rads). Remember the dose and time frame of radiation retinopathy with the **15-30 rhyme**.

**More than 15 months and at least 30 gray (3000 rads)
Typical before Retinopathay** (retinopathy)

Note: Sometimes as little as 15 Gray of radiation can cause radiation retinopathy. 1 gray = 100 rad

Information source: American Academy of Ophthalmology Basic and Clinical Science Course, Section 12, Retina and Vitreous 2008-2009, p. 178.

NOTES & YOUR OWN MNEMONICS:

NOTES & YOUR OWN MNEMONICS:

IX. STRABISMUS

SUBSIDIARY ACTIONS OF THE EXTRAOCULAR MUSCLES

The primary functions of the extraocular muscles are usually easy and straightforward to remember. Once we start talking about the subsidiary actions of the muscles (their secondary and tertiary functions) things get a bit harder. Grouping the muscles' functions together with this little mnemonic makes remembering subsidiary muscle actions a snap!

Remember – **SIN-RAD** for subsidiary muscle actions.

> The **S**uperior muscles **IN**tort,
> (therefore the inferior muscles must extort).
> The vertical **R**ecti **Ad**duct,
> (therefore the o**B**lique muscles must a**B**duct).

Mnemonic origin unknown. It may be found in Weinstock F, Contact Lens Fitting: A Clinical Text Atlas. Gower Medical Publishing, 1989.

MEASUREMENTS FOR THE SPIRAL OF TILLAUX

S_PI_R_AL_ of
TI ∠∠ AUX

5   5
6   5
6   6
7 . 7

The Spiral of Tillaux describes the distance from the limbus to muscle insertion for the four rectus muscles. The distances for the MR, IR, LR and SR respectively are 5.5 mm, 6.5 mm, 6.9 mm and 7.7 mm.

The Spiral of Tillaux is as easy as **5,6,7**. (See Illustration)

(For the lateral rectus measurement, the last six has to be flipped 180 degrees to become a 9.)

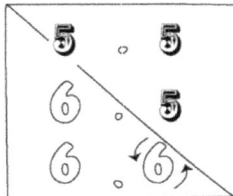

AND

The name **S**PIRAL OF TILLAUX gives you the **S**uperior **R**ectus measurement of **7.7** mm.

If you invert the "**LL**" of the word TILLAUX you get **7.7** which is the maximum distance of insertion in the **S**PIRAL OF TILLAUX (the insertion of the **S**uperior **R**ectus muscle at **7.7** mm).

OBLIQUE MUSCLES: ANGLE OF INSERTION

Both the superior oblique and the inferior oblique muscles insert at an angle 51 degrees from the direction of the visual axis when the eyes are in primary position. If you simply remember this mnemonic when you see the word "oblique," you'll never forget the insertion angle for the obliques.

The superior **obl**ique and inferior **obl**ique insert at **51** degrees from the visual axis.

oblique *becomes* o **51** ique *becomes* o **51** ique

If you simply add a horizontal stroke to the top of the "**b**" in the word "o**bl**ique" you will form the number **5**. The adjacent letter "**l**" looks just like the number **1**. And together the magical number **51** is formed, reminding you that both the o**bl**ique muscles insert at **51** degrees from the direction of the visual axis.

Information source: American Academy of Ophthalmology Basic and Clinical Science Course: Pediatric Ophthalmology & Strabismus. American Academy of Ophthalmology, 2008, p. 15.

HERING'S LAW AND SHERRINGTON'S LAW

Hering's Law of equal innervation states that there is equal innervation to the yoke muscles responsible for conjugate eye movements. Sherrington's Law of reciprocal innervation states that when an agonist muscle contracts, its antagonist relaxes to an equal extent. To keep these two laws organized in your mind remember:

Hering's Law makes the eyes move toget**Her**.*

That is to say that **Her**ing's Law applies to equal innervations of the yoke muscles allowing both eyes to move toget**Her** in a saccade.

AND

For "**Share**"ington's Law, the agonist/antagonist muscle pairs must **Share** available innervational impulses.

Thus, through the **Share**d innervations of "**Share**"ington's Law of reciprocal innervation, the agonist is stimulated to contract while the antagonist relaxes. For example, if the medial and lateral recti of the left eye **Share** innervation, when the medial rectus is stimulated to contract, it must take innervation away from the lateral rectus (after all, they **Share** the available innevational impulses). Likewise, if the lateral rectus is stimulated to contract, it must take innervation away from the medial rectus. Again, they **Share** the available innervational impulses.

*Mnemonic by permission of and modified from Steve Kraft, MD, FRCS(C).

SECONDARY DEVIATION VERSUS PRIMARY DEVIATION

When a patient presents with paralytic strabismus, the deviation can be measured and quantified either with the normal eye fixating or with the paretic eye fixating. The primary deviation is measured when the normal eye is fixating, and the secondary deviation is measured when the paretic eye is fixating. In paralytic strabismus, the secondary deviation (measured when the paretic eye fixating) is greater than the primary deviation.

Remember - **Two > One** for primary and secondary deviation measurements.

That is to say, **Two** (the secondary deviation) measures a greater deviation than **One** (the primary deviation) in paralytic strabismus.

CLASSIFICATION OF DUANE'S SYNDROME

There are three types of Duane's syndrome categorized by the patient's motility deficit. Patients with Duane's type I have poor abduction, type II have poor adduction, and type III have both poor abduction & adduction.

The number of "**D**'s" in the Deficit Determines the type of **D**uane's.

Poor ab**D**uction (1 "**D**") =   Type I Duane's
Poor a**DD**uction (2 "**D**'s") =   Type II Duane's
Poor ab**D**uction & a**DD**uction (3 "**D**'s") =Type III Duane's

Mnemonic origin unknown. This mnemonic is popular and I have heard it used several times in the U.S. and Canada.

ANOMALOUS RETINAL CORRESPONDENCE VERSUS ECCENTRIC FIXATION

Anomalous retinal correspondence (ARC) is a binocular phenomenon whereby two anomalous (disparate or different or non-corresponding) points on the retina come to be associated as the same point in space sensorially. In other words, these two non-corresponding points on the retina are treated by the visual system as though they actually do correspond. ARC cannot exist in a monocular patient. In fact, the words "retinal correspondence" in ARC refer to two corresponding retinal points (not just one point). Therefore two eyes must be involved, and ARC must be binocular.

Eccentric fixation, on the other hand, is a strictly monocular phenomenon whereby the patient fixates with a portion of the retina that is eccentric to the fovea. This may be observed in amblyopic eyes or eyes with retinal pathology.

Think – **B-ARC** when you see the word **ARC** .

Imagine that all children who develop **ARC** ("ark") suddenly drop to all fours and **B-ARC** ("bark") like a dog.

B-ARC tells you that **ARC** is a **B**ilateral phenomenon.

Eccentric fixati**ON** is a **ON**e eyed phenomenon.

ONE DELLE VERSUS MANY DELLEN

Dellen are a common corneal complication following strabismus surgery. To sound like a real surgeon, remember the difference between a delle (singular) and dellen (plural).

A del**LE** is sing**LE** (and they also rhyme).
On the other hand, delle**N** are **N**umerous.

LATERAL GENICULATE NUCLEUS ANATOMY: LATERALIZATION

Both eyes send visual information to both lateral geniculate nuclei (LGNs); however, each LGN only receives information from one side of the visual field (right or left visual field). Each of the six layers of the LGN receives this visual field information from either the ipsilateral or contralateral eye, depending on the layer. The eye on the same side as the LGN (the ipsilateral eye) sends information to layers 2, 3 and 5. The eye on the opposite side from the LGN (the contralateral eye) sends information to layers 1, 4 and 6.

For LGN anatomy remember - **See I? I see. I see.**

"**See**" stands for the "**C**" in **C**ontralateral.
"**I**" stands for the "**I**" in **I**psilateral.
In the order of the LGN layers 1 – 6.

Layer 1 = **C**ontralateral eye
Layer 2 = **I**psilateral eye
Layer 3 = **I**psilateral eye
Layer 4 = **C**ontralateral eye
Layer 5 = **I**psilateral eye
Layer 6 = **C**ontralateral eye

Mnemonic origin unknown, although it can be found at Wikipedia.com.

SURGICAL CORRECTION OF "A" AND "V" PATTERNS

During strabismus surgery, the horizontal rectus muscles can be vertically transposed to correct "A" or "V" pattern strabismus that is not caused by oblique overaction. Whether surgically correcting an "A" or "V" pattern, the medial recti should be displaced toward the apex of the letter and/or the lateral recti should be displaced toward the empty space in the letter. To remember this surgical strategy:

Think – **MALE** for the surgical correction of "A" or "V" pattern strabismus.

Displace **M**edial recti towards **A**pex of pattern and,
Lateral recti towards the **E**mpty space.

Information source: American Academy of Ophthalmology Basic and Clinical Science Course: Pediatric Ophthalmology & Strabismus. American Academy of Ophthalmology, 2008, p. 123.

CYCLOPLEGICS: DURATION OF ACTION

The cycloplegics vary in their duration of action from several hours to even one or two weeks. Atropine, commonly used for penalization in amblyopia, has the longest duration at 1-2 weeks while tropicamide, commonly used for dilation in the office, has the shortest duration of action at only 4-8 hours. The other agents are of intermediate duration of action as detailed below:

AGENT	DURATION OF ACTION
Atropine	1-2 weeks
Scopolamine	5-7 days
Homatropine	1-3 days
Cyclopentolate	8-24 hours
Tropicamide	4-8 hours

Remember - **ASH-CT** and **The Rule of 1/3's**
For the duration of action of the cyclopegics from longest to shortest.

ASH-CT

Atropine (longest)
Scopolamine
Homatropine
Cyclopentolate
Tropicamide (shortest)

And the Rule of 1/3's

Starting with Atropine, with a duration of action of approximately 14 days, each subsequent cycloplegic in the **ASH-CT** mnemonic has a duration of action that is **1/3** the previous cycloplegic.

Atropine	(approximately 14 days)
Scopolamine	(1/3 of 14 ≈ 5 days)
Homatropine	(1/3 of 5 days ≈ 1.5 days)
Cyclopentolate	(1/3 of 1.5 days ≈ 12 hours)
Tropicamide	(1/3 of 12 hours ≈ 4 hours)

ANGLE KAPPA	A positive angle kappa will create the appearance of an exotropia, when in fact, no deviation exists. Positive angle kappa can be due to conditions that drag the macula such as staphyloma or ROP.

Remember – A positive "**+**"ve angle kappa simulates an e**X**otropia.

The "**+**" sign of positive reminds you of the "**X**" in e**X**otropia.

AVERAGE NORMAL FUSIONAL AMPLITUDES

Fusional amplitudes (FA's) are a favorite question in the strabismus clinic, especially the difference between normal vertical and convergence fusional amplitudes. Using the **Rule of 2.5's** for normal fusional amplitudes detailed below, you can quickly construct a complete table of a normal vertical, divergence, and convergence fusional amplitudes both at distance and at near – all while remembering only one number – **2.5**!

The Rule of 2.5's for normal fusional amplitudes. (See Table)

Take a minute to quickly draw out this table, and you'll see how simple **The Rule of 2.5's** really is. The rule states:

1) **Vertical** FA's are **2.5** PD at distance and near. They are small and almost identical.
2) At a **Distance**:
        a) Divergence FA is **2.5** times the vertical FA
        b) Convergence FA is **2.5** times the divergence FA
3) At **Near**, both divergence and convergence FA's are **2.5** times the distance values for divergence and convergence FA's

If you have trouble constructing the table below, remember:
a) **DI**vergence FA's **DI**minish **2.5 X** compared to convergence FA's.
b) **DI**stance horizontal FA's **DI**minish **2.5 X** compared to those at near.

Test Distance	Vertical (Vert.) FA	Divergence Horiz. FA	Convergence Horiz. FA
6 m (20 ft) a.k.a. Distance			
ACTUAL	2.5	6	14
By the Rule of 2.5's	**2.5**	*2.5 x **2.5** = 6.25*	*6.25 x **2.5** = 15.6*
25 cm (10 in) a.k.a. Near			
ACTUAL	2.6	16	38
By the Rule of 2.5's	**2.5**	*6.25 x **2.5** = 15.6*	*15.6 x **2.5** = 39*

Note: The reference measurement for near in this table is **25 cm**, rather than 33 cm.

Information source: American Academy of Ophthalmology Basic and Clinical Science Course: Pediatric Ophthalmology & Strabismus. American Academy of Ophthalmology, 2008, p. 123.

PRISM MEASUREMENTS

To neutralize an exotropia, place the prism base in. To neutralize an esotropia, place the prism base out.

Think - **BIX-BOES** to neutralize a deviation.

Base-**I**n for e**X**otropia.
Base-**O**ut for **ES**otropia.

Mnemonic origin unknown.

THE COVER TESTS

Three common cover tests for assessing strabismus are the monocular cover-uncover test, the alternate cover test, and the simultaneous prism and cover test. Each test will demonstrate a different component of the deviation.

1) Movement on the monocular cover-uncover test indicates the presence of a tropia. The cover-uncover test only tells you that a tropia is present. It is not used to measure or quantify the degree of that tropia because no prisms are used.

Remember – **CUT** for the monocular **C**over-**U**ncover test for **T**ropias.

In other words, a deviation on the monocular **C**over-**U**ncover test indicates a **T**ropia.

2) The alternate cover test (a.k.a alternate prism and cover test) is used to quantify and measure the total deviation (BOTH the tropia component AND the phoria component, combined).

ALternating prism cover test measures tot**AL** deviation.

OR

ALTO – The **AL**ternate cover test measures **TO**tal deviation.

3) The simultaneous prism and cover test is used to isolate and measure ONLY the tropia component of a deviation. It is especially useful in the monofixation syndrome where patients may have a small tropia combined with a large phoria. In this case, the alternate cover test may show a large deviation (because of the large phoria). The simultaneous prism and cover test then be used to isolate ONLY the smaller tropia component of the deviation.

Sighing **mules tan** in the **tropi**cs.

The "**Sighmuletan**"enous (simultaneous) prism cover test measures only the **tropi**a component of a deviation.

101

4) Combine all of the cover tests into one supermnemonic.

Think – "**STrUTAITo**" to remember all three cover tests.

Simultaneous cover test	**Tr**opia (isolates and quantifies the Tropia)
Cover-**U**ncover test	**T**ropia (detects the presence of a Tropia)
Alternate cover test	**To**tal deviation (quantifies Total deviation = Tropia + Phoria)

Information source: American Academy of Ophthalmology Basic and Clinical Science Course: Pediatric Ophthalmology & Strabismus. American Academy of Ophthalmology, 2005. pp. 74-77.

THE 9 DOTS OF THE TITMUS STEREO TEST

When administering the Titmus stereo test utilizing the nine Titmus circles, it is important for you, the examiner to have the correct answers memorized (or written on the back of the book). Luckily, the 9 Titmus stereo circles are easy to remember with this mnemonic.

Remember – "**BLoB ToTaL RuLeR**" for the Titmus stereo circles. (See Illustration)

Imagine the stereo fly, big and fat, sitting like a **BLoB** atop the stereo test book and staking his claim as "**ToTaL RuLeR**" of the book.

B = **B**ottom   **T** = **T**op   **R** = **R**ight   **L** = **L**eft

BLoB	Row 1 = **B**ottom, **L**eft, **B**ottom
ToTaL	Row 2 = **T**op, **T**op, **L**eft
RuLeR	Row 3 = **R**ight, **L**eft, **R**ight

FISSURE NARROWING IN DUANE'S SYNDROME

In Duane's syndrome, the palpebral fissure narrows on attempted adduction of the affected eye. This is due to co-contraction of the medial and lateral recti on attempted adduction and causes a relative enophthalmos, thus narrowing of the palpebral fissure.

In **DuaNe**'s, **AD**duction **N**arrows the palpebral fissure.

MEASURING AC/A RATIO	The AC/A ratio in patients can be measured by either the lens gradient method or the heterophoria method. In the **lens gradient method**, changes in deviation at a <u>fixed distance</u> are measured <u>using different power lenses</u> (usually minus lenses) over the eyes. For the **heterophoria method**, changes in prismatic deviation are measured at <u>varying distances</u> (near and far) <u>without changing any lenses</u>. The interpupillary distance in centimeters (PDc) must be measured to calculate the AC/A ratio using the heterophoria method. In a normal patient, the AC/A ratio by the heterophoria method (called "heterophor" below) equals the PDc, but may be more depending on the disparity of near and distance measurements. Remember the difference in these two methods of calculating AC/A ratio with this little rhyme:

Gradient, many lens measurements;
Heterophor, is PDc & maybe more.

NYSTAGMUS: PRISM THERAPY	Nystagmus may be less pronounced when the patient's field of gaze is directed into the null zone of the patient's nystagmus. When encountered with a patient who has nystagmus with a null zone and compensatory face turn, you can place a prism with its base opposite to the direction of the null zone to abolish the face turn.

Remember - **BOp NoZe** to eliminate a face turn in nystagmus.

That is, place the prism **BOp NoZe** (**B**ase **Op**posite to the **N**ull **Z**one). Pretend you can fix the face turn by bopping the patient with nystagmus on the nose.

PARALYTIC STRABISMUS: EPONYMS FOR ABDUCENS PALSY SURGERY	In the case of paralytic strabismus due to abducens palsy, surgical correction is approached by trying to bolster or supplement the force of the paralyzed lateral rectus (LR) muscle. Both the Jensen procedure and the Hummelsheim procedure take advantage of the force of the superior rectus (SR) and inferior rectus (IR) to bolster the LR. In the Jensen procedure, the SR and IR are split (but NOT disinserted) and sutured to the split LR at their midportion at least 12 mm posterior to the insertion. In the Hummelsheim procedure, the SR and IR are split and the split ends ARE disinserted and sutured adjacent to the LR insertion. Remember the difference as follows:

Jensen **J**oins and **H**ummelsheim **H**alves.

For the **J**ensen procedure, simply **J**oin the split portions of the LR, SR, and IR at their midportion (no disinsertion or transposition required).

For the **H**ummelsheim procedure, **H**alf the SR and IR, disinsert the SR and IR **H**alves, and transpose those **H**alves to the LR insertion (This is a **H**alf tendon transposition not full tendon transfer).

Mnemonic origin unknown. Popular in Canada and the U.S.

MAGNOCELLULAR AND PARVOCELLULAR NEUROPHYSIOLOGY

The magnocellular neurons (a.k.a. M cells or large cells) are used for determining *where* things are. From the striate cortex, information from M cells goes predominantly to parieto-occipital areas. On the other hand, parvocellular neurons (a.k.a. P cells or small cells) are used to examine static objects and determine *what* things are. From the striate cortex, information from P cells goes predominantly to the temporal-occipital areas. To remember these functions and associations of neurons:

Imagine a golfer **Wear**ing only a **MOP** and on the putting green. He asks "**What**'s **Par** on this hole?" just before he **POT**s (sinks) the ball.

Wearing a **MOP** reminds you that **Where** things are is determined by the **M**agnocellular cells to the **O**ccipital-**P**arietal area.

"**What**'s **Par**" as he **POT**s the ball reminds you that **What** things are is determined by the **Par**vocellular cells to the **O**ccipital-**Te**mporal area.

Information source: American Academy of Ophthalmology Basic and Clinical Science Course: Pediatric Ophthalmology & Strabismus. American Academy of Ophthalmology, 2008, pp. 44-46.

NOTES & YOUR OWN MNEMONICS:

NOTES & YOUR OWN MNEMONICS:

X. PEDIATRICS

OCULAR EMBRYONIC DEVELOPMENT

The embryonic eye develops from the folds of the neural plate to become the optic pit, then the optic vesicle, and then the optic cup.

Think – **PVC** for the order of ocular embryogenesis.

PVC is **P**it, then **V**esicle, then **C**up.
The period of embryonic development is one of neural plasticity just like **PVC** (Polyvinyl chloride) is a type of plastic.

CLOSURE OF THE EMBRYONIC FISSURE

The embryonic fissure of the optic cup closes at day 33. Incomplete closure results in colobomas.

Closure of the embryonic FI**33**URE occurs at day **33**.

The two "**S**'s" in the word "FISSURE" can easily be made to resemble "**3**'s" to form the word "FI**33**URE." Make the "**S**'s" into "**3**'s" every time you see the word "fissure," and you'll never forget the date of closure of the embryonic FI**33**URE – day **33**!

CORNEAL CLOUDING IN INFANCY

Corneal clouding in infancy is often associated with congenital glaucoma. Other causes to include in your differential diagnosis include sclerocornea, congenital ichthyosis, forceps birth trauma, corneal ulcer, metabolic disorders such as mucopolysaccharidoses or mucolipidoses, Peter's anomaly, posterior keratoconus, staphyloma, congenital hereditary endothelial dystrophy (CHED), posterior polymorphous corneal dystrophy (PPMD), and a central corneal dermoid.

Remember – **STUMPED** for the differential of corneal clouding in infancy.

Sclerocornea, **S**kin problems (e.g. congenital ichthyosis)
Tears in Descemet's (from Glaucoma or forceps birth trauma)
Ulcer
Metabolic diseases (e.g. mucopolysaccharidoses or mucolipidoses, although these conditions rarely present at birth)
Posterior corneal defect (e.g. **P**eter's anomaly, **P**osterior keratoconus, or Staphyloma)
Endothelial dystrophy (CHED & PPMD)
Dermoid (e.g. central corneal dermoid)

Mnemonic by permission of and modified from George O. Waring, III, MD as appeared in J. Judge, G. Waring III, & R. Blocker, H. Leibowitz & G. Waring III (eds), Congenital and Neonatal Corneal Abnormalities in Corneal Disorders: Clinical Diagnosis and Management. W. B. Saunders, 1998, p. 208.

THE VACANT MUCOPOLYSACCHARIDOSIS SYNDROME

The mucopolysaccharidoses are numbered from I to VII. However, MPS V is vacant (does not exist), a favorite trivia question.

Think - **MPS Five is Not Alive.**   OR   MPS **V** is **V**acant.

Whether you remember the rhyme reminiscent of the fruit drink trade named "Five Alive" or the association with the Roman numeral "**V**" and **V**acant, either way, MPS V does not exist.

INBORN ERRORS OF METABOLISM	The vast majority of the inborn errors of metabolism are autosomal recessive; a few are X-linked recessive. Remember that X-linked recessive disorders predominantly occur in boys since males have only one X-chromosome. When confronted with a question about the inheritance of an inborn error of metabolism, remember the majority are autosomal recessive, the exceptions are X-linked recessive. Two of the exceptions (X-linked recessive metabolic disorders) are Hunter's disease and Fabry's disease.

For **Hunt**er's disease, remember - Most people who **Hunt** are men. Thus **Hunt**er's disease occurs in boys & is X-linked recessive. In fact, **Hunt**er's disease is the ONLY mucopolysaccharidosis that is X-linked recessive.

For Fabry's disease - Think Fab**Ray**'s disease. **Ray** is a boy's name, Fab**Ray**'s occurs in boys & is X-linked recessive. In fact, Fab**Ray**'s disease is the ONLY lipidosis that is X-linked recessive.

MUCOPOLYSACCHARIDOSES WHERE THE CORNEA IS USUALLY CLEAR	The majority of the mucopolysaccharidoses result in corneal clouding. However, the two exceptions to this rule are Hunter's (mucopolyscaccharidosis type II) and Sanfillipo's (mucopolyscaccharidosis type III) which usually have clear corneas. Remember these two mucopolysaccharidosis exceptions with this little rhyme:

**In Hunter's and Sanfillip's,
The cornea's not opique.**

That is to say, the two mucopolysaccharidoses with clear corneas are **Hunter's** and Sanfillipo's ("**Sanfallip's**") where the cornea is not opaque ("**opique**").

Note: Some cases of Hunter's syndrome may show mild corneal clouding.

GALACTOSEMIA	Galactosemia may be caused by defects in three different enzymes: galactose-1-phosphate uridyl transferase, galactokinase, or UDP galactose epimerase. Infants with any of these enzyme deficiencies will develop cataract. Galactose-1-phosphate uridyl transferase deficiency is the most common enzyme defect causing galactosemia in infants. In addition to developing cataracts, these infants are usually systemically ill (multi-organ dysfunction, anemia, and failure to thrive). The galactokinase and epimerase deficiencies are less common, and the children have less severe systemic manifestations.

For galactosemia, a deficiency of **Transf**erase is **Tri**te and **Transf**orms a baby more.

Trite means "commonplace," and **Transf**erase deficiency is the most common enzyme deficiency in galactosemia. "**Trans**forms a baby more" reminds you that the **Trans**ferase deficiency causes more systemic abnormalities than the galactokinase or epimerase defects.

Information source: American Academy of Ophthalmology Basic and Clinical Science Course: Lens and Cataract. American Academy of Ophthalmology, 2008, pp. 60-61.

BIOCHEMICAL DEFECT IN TAY-SACH'S DISEASE

Tay Sach's disease is a lysosomal storage disease (mucopolysaccharidosis type 1, a GM2 **Ganglio**sidosis). The enzyme defect in Tay-Sach's disease absent hexosaminidase A. GM2 **Ganglio**sides are deposited in the **Ganglio**n cell layer, which is absent in the fovea. For this reason, a cherry red spot is seen in Tay-Sach's disease because of increased contrast of the choroid underlying the fovea (WITHOUT an opaque **Ganglio**side-packed **Ganglio**n cell layer) versus the rest of the retina (WITH an opaque **Ganglio**side-packed **Ganglio**n cell layer). Remember these details of Tay-Sach's disease with this little story:

Imagine a wicked witch with a tall "**A**"-shaped hat casting **hexes** by throwing **cherries** at unfortunate kids, giving them Tay Sach's disease. To stop her, **two GM** trucks **gang-up** on the witch and **Tie** her up in a **Sack**. (See Illustration)

The "**A**"-shaped hat and casting **hexes** remind you of **hex**osaminidase **A** deficiency causing Tay-Sach's.
By throwing **cherries** reminds you of the **cherry-red spot** of Tay-Sach's disease.
The **two GM** trucks that **gang-up** on her remind you of **GM2 Gangl**iosidosis.
And **Tie** her up in a **Sack** simply reminds you of the name "**Tay Sach's.**"

ANIRIDIA & WILMS' TUMOR

Wilms' tumor is associated with sporadic aniridia and almost never with familial aniridia. In fact, up to 1/3 of patients with sporadic aniridia will develop Wilms' tumor. The chromosomal defect in causing aniridia (whether familial or sporadic) is located on chromosome 11p13. Here are few mnemonics to keep all this straight.

Think - **Spor**adic aniridia has tumor "**Spor**es" causing Wilms' tumor. While **Famili**al aniridia "protects" children against Wilms' tumors just as **Famili**es protect their children.

Familial or sporadic aniridee,
Is caused by mutations at eleven p one three.

This little rhyme reminds you familial or sporadic aniridia (**aniridee**), is caused by a chromosomal mutation at the 11p13 (pronounced "**eleven p one three**") locus.

Just as mutations at eleven p **one three** cause aniridia;
So too, **one in three** patients with **Spor**adic aniridia "**Spor**es" will develop Wilms' tumor.

Information source: American Academy of Ophthalmology Basic and Clinical Science Course: Pediatric Ophthalmology & Strabismus. American Academy of Ophthalmology, 1994, p. 69.

DIFFERENTIAL DIAGNOSIS FOR ECTOPIA LENTIS

Ocular conditions commonly associated with ectopia lentis are ocular trauma, aniridia, simple ectopia lentis, and ectopia lentis et pupillae. Systemic disorders associated with ectopia lentis include Marfan syndrome, homocystinuria, Weill-Marchesani syndrome, hyperlysinemia, sulfite oxidase deficiency, syphilis, and Ehlers-Danlos syndrome. To remember the differential for ectopia lentis:

mnemonic

Think - "**We'll MES$_2$H$_2$ TAPES**" to fix the ectopia lentis.
(See Illustration)
Imagine sitting in the O.R. and deciding to use a meshwork of tape to re-suspend the lens. You look at the scrub nurse and say, "**We'll MES$_2$H$_2$ TAPES**" to fix this ectopia lentis.

We'll MESH = Weill-Marchesani

Marfans	**T**rauma
Ehler-Danlos	**A**niridia
Syphilis, **S**ulfite oxidase deficiency	**P**upillae (et **P**upillae)
Homocystinuria, **H**yperlysinemia	**E**ctopia lentis
	Simple

Information source: American Academy of Ophthalmology Basic and Clinical Science Course: Pediatric Ophthalmology & Strabismus. American Academy of Ophthalmology, 2008, p. 305.

DIRECTION OF LENS DISLOCATION IN ECTOPIA LENTIS

In ectopia lentis due to Marfan syndrome, the lens dislocates superiorly or temporally. In contrast, in ectopia lentis associated with homocystinuria, the lens tends to dislocate inferiorly.

mnemonic

Marfan sounds like **Martian**.

Martians are found **up and out** in space,
Just like the lens tends to dislocate **up and out** in **Marfan** syndrome.

mnemonic

In homocyst**IN**uria, the lens dislocates **IN**ferior.

OR

mnemonic

Remember - The **Cyst** with **Urine** in it (a.k.a. the bladder) is located **inferior**ly.

Thus homo**Cyst**in**Uria** is associated with **inferior** lens dislocation.

OPSOCLONUS: PEDIATRIC

Opsoclonus in children is commonly associated with neuroblastoma. I used to confuse this with medulloblastoma until I thought of a pediatric neuro-ophthalmologist saying:

"Pediatric NEURO-**op**hthalmologists are a BLAST **so clone us**!"

This reminds you that in the Pediatric population NEURO-ELAST-oma (neuroblastoma) is associated with **op-so-clone-us** (opsoclonus).

OPTIC DISC HYPOPLASIA

Optic disc hypoplasia may be associated with maternal diabetes, maternal phenytoin (trade name Dilantin) ingestion, or maternal alcohol ingestion. The condition can be isolated or associated with De Morsier's syndrome (septo-optic dysplasia with possible pituitary hormone deficiency), anterior visual pathway tumors, or other CNS developmental abnormalities such as holoprosencephaly, lissencephaly, or porencephaly. On exam the optic disc in optic disc hypoplasia may have a "double-ring" appearance as you observe the scleral ring surrounding the small, hypoplastic ring of the optic disc.

Remember - The **7 D's** of optic **D**isc hypoplasia.

Double-ring sign
Diabetes (maternal diabetes)
Dilantin (maternal ingestion of phenytoin (trade name **D**ilantin))
Drinking alcohol (maternal ingestion of alcohol)
De Morsier's syndrome (septo-optic dysplasia)
Developmental CNS abnormalities (e.g. holopros-, liss-, & porencephaly)
Distensions (anterior visual pathway tumors)

Mnemonic by permission of and modified from Rosa Tang, MD.

Information source: David Taylor & Creig S. Hoyt (eds), Pediatric Ophthalmology and Strabismus. Elsevier Saunders, 2005, p. 625.

CONGENITAL OPTIC NERVE PITS

Congenital optic nerve pits may be incidental exam findings or associated with decreased vision due to serous retinopathy. These pits are usually located on the inferior temporal aspect of the optic disc.

Remember – The **PITS** of optic nerve pits.

Pits (optic nerve pits)
Inferior **T**emporal location
Serous retinopathy

Congenital optic nerve **P**its are usually located on the **I**nferior **T**emporal optic disc, and may be associated with **S**erous retinopathy due to serous fluid leakage from the pit into the subretinal space.

HAMARTOMA VERSUS CHORISTOMA	A hamartoma is an anomalous tissue formation consisting of only those elements that ARE normally found at the involved site. Conversely, a choristoma is an anomalous tissue formation consisting of tissue elements that ARE NOT normally present at the involved site.

Remember - **HA**martomas consists of tissue elements **HA**bitually found at the involved site.

By exclusion, choristomas consist of tissue elements NOT habitually found at the site.

LEUKOCORIA	For many, leukocoria immediately brings to mind the possibility of the retinoblastoma or juvenile cataract; however, not all leukocoria is RB or cataract. Use this mnemonic to carry you through the broad differential diagnosis.

Remember - **RCMP** for the differential of leukocoria.

First of all, **RCMP** is the abbreviation for the **R**oyal **C**anadian **M**ounted **P**olice. Imagine, a **R**oyal **C**anadian **M**ounted **P**oliceman (**RCMP**) sitting on his horse atop a snowy white retinoblastoma tumor. Because this differential is long, it may be useful to remember there are 7 "**R's**," 6 "**C's**," 4 "**M's**," and 3 "**P's**."

Retinoblastoma, **R**etinopathy of Prematurity, **R**etinal Folds (congenital), **R**hegmatogenous Retinal Detachment, **R**etinal Dysplasia (recessive), **R**etinoschisis (juvenile X-linked), and **R**etinal tumor (other retinal tumors including: astrocytic hamartoma or hemangioblastoma)

Coat's disease, **C**ariasis (toxocariasis), **C**ataract, **C**oloboma, **C**horoidal hemangioma, and **C**horioretinitis (uveitis)

Medulloepithelioma, **M**yelinated nerve fibers, **M**etastatic disease (endophthalmitis or leukemic mets), and the **M**artoma's (astrocytic hamartoma* or combined retina and RPE hamartoma)

PHPV, **P**rematurity (if you forgot ROP), and **P**lasmosis (Toxoplasmosis)

*Note: Astrocytic hamartoma is repeated twice in case you forget.

TREATMENT OF KAWASAKI SYNDROME	Kawasaki syndrome (a.k.a. the mucocutaneous lymph node syndrome) may present to the ophthalmologist as bilateral conjunctival injection in a young child. A major cause of morbidity and even death in these patients is coronary artery aneurysm, which is worsened by the use of systemic steroids. Aspirin is the drug of choice for Kawasaki syndrome.

In Kaw**ASA**ki's syndrome, **ASA** is the drug of choice.

Conveniently, the name "Kaw**ASA**ki's" has the abbreviation for aspirin (**ASA**) right in the name.

MORNING GLORY DISC ANOMALY

The morning glory disc anomaly can be associated with basal encephalocele (abbreviated **BE** in the mnemonic below). An extreme and uncommon form of basal encephalocele is a transphenoidal encephalocele which may present as nasal obstruction, rhinorrhea, mouth breathing, snoring, or pulsatile "nasal polyps" that should not be biopsied.

Morning glory disc anomaly is rarely familial, and rarely bilateral. The optic disc in morning glory syndrome has abnormal retinal vessels that appear increased in number and appear to emerge from the disc in a straightened radial fashion (like spokes of a wheel). The anomaly is usually unilateral and more frequent in females. The acuity in the affected eye is in the 20/200 to finger counting range. Remember the associations of morning glory disc anomaly with this little rhyme to the tune of the doxology "Glory Be" (a.k.a. "Glory Be to the Father").

Morning Glory **BE!**
Happens sporadically,
In one eye of girls, most frequently.

Morning Glory **BE!**
Disc in depression centrally,
Central tuft you'll see,
With Straight vessels Radially.

That is to say, Morning Glory disc anomaly is associated with **BE** (**B**asal **E**ncephalocele), and is usually sporadic, unilateral, and found in females. The disc in Morning Glory anomaly will appear centrally depressed with a central tuft and straight, radial vessels extending like spokes on a wheel.

Information source: Brodsky, Baker & Hamed, Pediatric Neuro-ophthalmology. Springer-Verlag, 1996, pp. 54-55.

CLINICAL CHARACTERISTICS OF CONGENITAL RETINAL DYSTROPHY

In clinic, if you see a patient with congenital nystagmus, light aversion with severe photophobia, high myopia, the oculodigital sign, and paradoxical pupils, you should request an electroretinogram (ERG) looking for congenital retinal dystrophy. Remember these signs and symptoms of congenital retinal dystrophy with this little rhyme:

Order an **ERG** for "**LMNOP**."

ERG reminds you to order an **E**lectro**R**etino**G**ram for the signs below.

Light aversion with severe photophobia
Myopia (high myopia)
Nystagmus (congenital nystagmus)
Oculodigital reflex
Paradoxical **P**upils

Information source: Brodsky, Baker & Hamed, Pediatric Neuro-ophthalmology. Springer-Verlag, 1996, pp. 306-307.

| **FINDINGS IN AICARDI SYNDROME** | Aicardi syndrome is an X-linked dominant disease seen in females that is characterized by agenesis of the corpus callosum, chorioretinal lacunae, colobomas of the optic nerve, mental retardation, and infantile spasms. |

Remember the word **AICARdI'S** for the signs of Aicardi syndrome.

Agenesis of corpus callosum
Infantile spasms
Chorioretinal lacunae, **C**olobomas of optic nerve
Agenesis of the corpus callosum (the "A" is a repeat of above)
Retar**d**ation (mental retardation)
Infantile **S**pasms (the "I" is a repeat of above also)

AICARDI'S

If you stretch your imagination you can find the letter "X" within the capital letter "R" of **AICARdI'S**. The distorted X much resembles the X chromosome and reminds you that **AICARdI'S** is an X-linked dominant (XD) disease.

| **ROP THRESHOLD CRITERIA** | Memorizing the criteria for threshold disease in ROP can be confusing with all the numbers, zones, and clock hours. For this mnemonic, I'll simplify everything down to make it as easy as 1, 2, 3. In brief, treatment of ROP is recommended for disease in zone 1 or zone 2, of stage 3+, with 5 contiguous clock hours of involvement or 8 total clock hours. To get these numbers as easy as 1, 2, 3: |

Treatment threshold for ROP is as easy as **1, 2, 3**.
1 + 2 = 3
  2 + 3 = **5**
    3 + 5 = **8**

Take the numbers 1 and 2, add them, and get the sum 3. Then add the 3 to the previous number, 2, to get the sum 5. Finally add the 5 to the previous number, 3, to get the sum 8. Now you have all the numbers important for the treatment threshold for ROP: 1, 2, 3, 5, & 8.

That's disease in zone **1** or zone **2**, of stage **3**+, with **5** contiguous clock hours of involvement or **8** total clock hours.

| **TRILATERAL RETINOBLASTOMA** | Trilateral retinoblastoma is defined as bilateral retinoblastoma associated with ectopic intracranial retinoblastoma, usually in the pineal gland or the parasellar region. Trilateral retinoblastoma is thought to occur in 3% of cases of retinoblastoma. |

TRIlateral retinoblastoma occurs in **3**% of retinoblastoma cases.

Information source: Carol L. Shields, James Vander & Janice Gault (eds), Retinoblastoma in Ophthalmology Secrets. Mosby, 2007, p. 406.

ALBINISM

1) There are various forms of albinism, and each has its own inheritance pattern. Oculocutaneous albinism is autosomal recessive, ocular albinism (for all practical purposes) is X-linked recessive, and ocular albinoidism is autosomal dominant.

Even **O**s**CAR** on Sesame Street knows **O**culo**C**utaneous albinism is **A**utosomal **R**ecessive.

You only have to be as smart as an **OX** to know that **O**cular albinism is **X**-linked recessive.

Ocular albinoi**D**ism is autosomal **D**ominant.

OR   Alternatively, you can combine the three mnemonics into one albinism inheritance supermnemonic:

Os**CAR** the **OX** is albinoi**D**.

Oculo**C**utaneous albinism is **A**utosomal **R**ecessive
Ocular albinism is **X**-linked recessive
Ocular albinoi**D**ism is autosomal **D**ominant

2) Patients with ocular albinism have increased decussation of temporal optic nerve fibers in the chiasm compared to normals, and this is now a diagnostic criterion for albinism.

Ocular albinos have more **crossed eyes** and **crossed fibers** than normals.

That is to say, patients with ocular albinism have and increased incidence of strabismus (**crossed eyes**) and increased decussation of temporal optic nerve fibers (**crossed fibers**) than normal patients.

3) Oculocutaneous albinism can be part of systemic syndrome that includes hematologic abnormalities. Oculocutaneous albinism associated with bleeding diathesis is Hermansky-**Pud**lak syndrome, while oculocutaneous albinism associated with white blood cell dysfunction and multiple infections is **Ch**édiak-Higa**shi** syndrome. Remember these associations with the following:

Imagine you patient standing in a **Pud**dle of blood if you miss the diagnosis of Hermansky-**Pud**lak syndrome with bleeding diathesis.

The name "**Ch**édiak-Higa**shi**" sounds like a sneeze, "**ACH**oo!" and reminds you of the frequent infections these children with white blood cell dysfunction are prone to.

| GENETICS OF NEUROFIBROMATOSIS | Neurofibromatosis type 1 (NF type 1) is associated with Lisch nodules and is due to a defect on chromosome 17. On the other hand, NF type 2 does NOT have Lisch nodules and is associated with a defect on chromosome 22. |

"**LI**"SCH nodules occur in NF type **1** due to a defect on chromosome "**17**."

The word "**LISCH**" has the number "**17**" right in it. Simply capitalize the word "**LISCH**" and rotate the "**LI**" 180 degrees to form the number "**17**." The number "**1**" in the number "**17**" tells you that Lisch nodules are associated with NF type **1**, and the **17** localizes the chromosome defect.

NF type **2** is due to a defect on chromosome **22** (and is not associated with Lisch nodules).

POINTS ON VON HIPPEL-LINDAU DISEASE

Von Hippel-Lindau disease (abbreviated VHL) is due to a defect on chromosome 3. The two major causes of death in VHL are renal cell carcinoma and cerebellar hemangioblastoma.

"**VHL**" has **3 letters** and VHL is due to a defect of chromosome **3**.

**VHL gives kidneys hell,
And isn't good for the cerebell.**

This little rhyme reminds you that the two main causes of death in Von Hippel-Lindau (**VHL**) disease are renal carcinoma (**gives the kidneys hell**) and cerebellar hemangioblastoma (**and isn't good for the cerebell**).

ATAXIA-TELANGIECTASIA: EPONYM & INHERITANCE

Ataxia-telangiectasia is also called Louis Barr syndrome. The gene for this autosomal recessive disease has been found on the long arm of chromosome 11 (11q22-23). It results in ataxia, cutaneous and retinal telangiectasias, immunodeficiency with thymic aplasia, and neuro-degeneration.

After drinking at **Louis' BAR** you'll be **ataxic**.
And Louis' B**AR** (Louis BA**R**r syndrome) is **A**utosomal **R**ecessive.

Ataxia **T**elangiectasia = **A**bsent/**A**bnormal **T**hymus

The "*ataQxia te11angiectasia*" locus in on chromosome **11q**.

When you see the words "ataxia telangiectasia", spell them with a "q" instead of an "x" and use two "1's" in the place of the "l." This gives you "*ataQxia te11angiectasia*" and will te**11** you the answer! The "*ataQxia te11angiectasia*" locus is on chromosome **11q**.

Note: The small letter "q" denotes the long arm of the chromosome. It is capitalized in the mnemonic for emphasis.

Mnemonic origin unknown. Special thanks to Mike Rauser M.D. for mentioning it.

TUBEROUS SCLEROSIS

Tuberous sclerosis can be due to a genetic defect in a tumor suppressor gene on either chromosome 9 or chromosome 16. The classic manifestations of tuberous sclerosis are adenoma sebaceum (angiofibroma), ash leaf depigmented skin lesions, seizures, and mental retardation. These patients may have subpleural cysts and develop spontaneous pneumothorax. You can remember all of these characteristics of tuberous sclerosis with the following mnemonics:

Imagine a **tube-faced** boy who just turned **sweet 16**. **Nine cats with 9 lives** find him so sweet they try to eat him and claw at his tube-shaped face leaving many **facial blemishes**. The boy tries to get rid of the cats by **shaking them off**. The boy can't run away from the 9 cats because there are so many that they **retard his progress**. He wears a **strategically positioned leaf** to protect his anatomy from the 9 cats. (See Illustration)

Tube-faced (tuberous sclerosis)
Sweet 16 (chromosome 16)
Nine cats with 9 lives (chromosome 9)
Facial blemishes (adenoma sebaceum)
Shaking them off (seizure disorder)
Retard his progress (mental retardation)
Strategically positioned leaf (ash leaf spots)

AND

Patients with **tube**rous sclerosis may require a chest **tube** due to subpleural cysts.

116

NIH DIAGNOSTIC CRITERIA FOR NEUROFIBROMATOSIS TYPE 1 (NF1)

The diagnostic criteria for neurofibromatosis type 1 are met in an individual if two or more of these features are present.

1) Six or more café au lait macules over 5 mm in greatest diameter in prepubertal individuals and over 15 mm in greatest diameter in postpubertal individuals

2) Two or more neurofibromas of any type or one plexiform neurofibroma

3) Freckling in the axillary or inguinal regions (Crowe´s sign)

4) Optic glioma

5) Two or more Lisch nodules (iris harmartomas)

6) A distinctive osseous lesion such as sphenoid dysplasia or thinning of long bone cortex with or without pseudoarthrosis

7) A first-degree relative (parent, sibling, or offspring) with NF1 by the above criteria

mnemonic

Imagine you have **two or more**, **bony**, horribly **freckled relatives** who gave you the "**first degree**" when you were younger. Your relatives are **eyeing** some **lumpy N-OG** (eggnog) and **coffee** at a very cheap price – **six cups for a dime**. Pretend it is their habit of drinking lumpy **N-OG** and **coffee** that causes them to have abnormal bones, ugly armpit and groin freckles, Lisch iris lumps, proptosis (optic nerve glioma), and skin neurofibromas.

Two or more – Patients must meet **two or more** criteria

Bony – Distinctive **bone** lesions
Freckled – Axillary/inguinal **freckles**
Relatives & "first degree" – Any first-degree **relative** diagnosed with NF1
Eyeing & lumpy – Lisch nodules
N-OG – **N**eurofibroma(s) & **O**ptic **G**lioma
Coffee & six cups for a dime – **Six café** au lait spots the size of a dime - post-pubertal (a dime is approximately 15 mm in diameter)

SPASMUS NUTANS

The findings in spasmus nutans include head nodding, torticollis, and nystagmus. It usually occurs within the first year of life. Spasmus nutans is usually benign but may be associated with chiasmal glioma or third ventricle tumors.

mnemonic

Remember - The **N**'s of **N**utans.

Nodding (head nodding)
Neck dystonia (torticollis)
Nystagmus
Neonatal onset
Neoplasm occasionally (including chiasmal glioma or third ventricle tumors)

CONGENITAL RETINAL DYSTROPHIES

The more common congenital retinal dystrophies include Leber's congenital amaurosis, congenital stationary night blindness, cone-rod dystrophy, rod-cone dystrophy, and achromatopsia (a partial form of achromatopsia is blue cone monochromatism).

Remember **LC₂A** for the congenital retinal dystrophies.

LCA is not only the abbreviation for **L**eber's **C**ongenital **A**maurosis But also for the more-common congenital retinal dystrophies!

Leber's congenital amaurosis
Congenital stationary night blindness
Cone-rod dystrophy & rod-cone dystrophy
Achromatopsia

CONGENITAL STATIONARY NIGHT BLINDNESS (CSNB)

On electroretinography, most patients with autosomal recessive and X-linked CSNB have a near-normal a-wave and a substantially reduced b-wave, in other words, an electronegative ERG.

CSNb stands for "**C**an't **S**ee **N**ormal **b** wave."

Imagine that you are the child's ophthalmologist. Looking at the patient's ERG the a-wave looks fine but the b-wave is gone. It's an electronegative ERG and because you "**C**an't **S**ee the **N**ormal **b** wave," you make the diagnosis – **CSNb**!

NOTES & YOUR OWN MNEMONICS:

NOTES & YOUR OWN MNEMONICS:

XI. PATHOLOGY

TYPES OF GIANT CELLS

There are three types of giant cells: Touton, foreign body, and Langhans. To remember the three types of giant cells:

mnemonic Imagine **Three Giants** weighing **Two Tons**, with **Forum**-sized **Bodies**, and **Long Hands**.

Three Giants (three types of giant cells)
weighing **Two Tons** (Touton giant cells)
with **Forum**-sized **Bodies** (Foreign body giant cells)
and **Long Hands** (Langhans giant cells)

LANGHANS CELLS VERSUS LANGERHANS CELLS

Langhans cells are a type of giant cell, and are classically associated with tuberculosis and sarcoidosis. Langerhans cells are seen in Langerhans histiocytosis (formerly known as Histiocytosis X). On electron microscopy, tennis racquet-shaped Birbeck granules are seen in Langerhans cells.

mnemonic Every time you see the word "Langerhans," put an "**X**" over the "**er**" in the word and replace it with an "**ir**."

Think of the "**er**" in the word "Langerhans" as an **er**ror in the word. So just take out your pen, and put a big, fat "**X**" right over that **er**ror. This reminds you that Langerhans cells are found in Histiocytosis **X**. You replaced the "**er**" with an "**ir**" because the "**ir**" reminds you of the "**ir**" in B**ir**beck granules found in Langerhans cells.

HISTOLOGIC APPEARANCE OF LANGHANS GIANT CELLS

Langhans giant cells are multinucleated with their nuclei in a "U" or horseshoe shaped arrangement.

mnemonic Imagine the finger tips of the **Long Handed Giants** mentioned above standing with their fingers crossed in front of their body forming a "**U**" shape.

OR

mnemonic Imagine person having such **long hands** that they can cheat at the game of **horseshoes** reaching all the way to the other side.

Either way, **Langhans** giant cells have "**U**" or **horseshoe** shaped nuclei.

APPEARANCE OF TOUTON GIANT CELLS

Touton giant cells (found in juvenile xanthogranuloma) have a ring of nuclei surrounding a central core of cytoplasm. The cytoplasm outside the ring of nuclei is typically foamy and vacuolated.

mnemonic **Touton** cells have **two-toned*** cytoplasm divided by a ring of nuclei.

AND

mnemonic Touton giant cells have a **C**entral **C**ytoplasm **C**ore with **O**uter vacu**O**lation.

*Mnemonic by permission of and modified from Robert Folberg, MD.

DISEASE ASSOCIATIONS WITH LANGHANS GIANT CELLS	Langhans giant cells are classically associated with tuberculosis and sarcoidosis. If you can't remember this by rote, the following is a long-winded visual mnemonic. Most of us can remember that mycobacterium tuberculosis appears histologically as "red snappers" on acid fast stain. Also we all know that elevated angiotensin converting enzyme (ACE) is associated with sarcoidosis.

Picture a flying **ACE** in a biplane circling the **Long-Handed Giant** like in the movie "King Kong." As the giant coughs, **Red Snappers** are expelled from his tuberculous lungs, as the **Long Hands** move to cover the mouth.

The flying **ACE** and the **Long-Handed Giant** remind you that **Langhans** cells are found in sarcoidosis with an elevated **ACE** level.

The **Red Snappers** and **Long Hands** covering the mouth remind you that **Langhans** cells are also found in tuberculosis which appears as **Red Snappers** on acid fast stain.

MERKEL CELL TUMORS	Merkel cell tumors are rare, usually purple, lid lesions that just keep showing up on exams over and over. They appear as violaceous nodular lesions in older patients (different than the one shown in Yanoff's atlas). They are cutaneous apudomas, with carcinoid-like histology, and have up to a 20% associated mortality. I have personally been asked to identify this lesion four separate times – at rounds, and in oral and written exams. Remember Merkel cell tumors with this little rhyme:

The purple Merkel
Shows up like Steve Urkel.

That is, **Merkel** cell tumors are **purple** tumors that tend to show up at bad times such as exams. Just like the television character **Steve Urkel** in the syndicated show "Family Matters," **Merkel** cell carcinoma is not a pleasant thing to have around – especially with a 20% mortality rate. You want the dangerous and annoying **purple Merkel** to go away like the annoying Steve Urkel.

DYSKERATOSIS VERSUS PARAKERATOSIS

Dyskeratosis is individual cell keratinization within the lower levels of the epidermis. Parakeratosis is thickening of upper keratin layer with retention of nuclei in the keratin layer.

Imagine a **Dish** with a **Pear** lying on top of it. **Carrots** have been stabbed **in** the **Dish** and the **Pear**. Inside the **Pear** sits a **Nuclear** bomb. (See Illustration)

The **Dish** and the **Pear** with **carrots in** them remind you that both **Dish-carrotosis** (dyskeratosis) **and Pear-carrotosis** (parakeratosis) involve excess **carrot-in** (keratin) formation. The **Pear** sits on top of and above the **Dish**, reminding you that **Pear-carrotosis** involves thickening of the upper keratin layer of the epidermis while **Dish-carrotosis** involves individual cell keratinization of the lower, deeper layers of the epidermis. The **Nuclear** bomb sitting inside the **Pear** reminds you that **Pear-carrotosis** has the retention of **Nuclei** in the upper keratin layer of the epidermis.

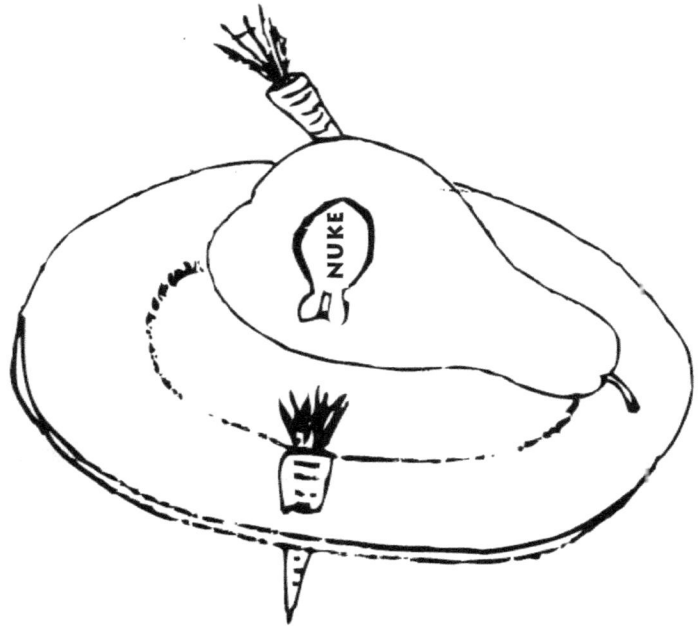

CONFIGURATION OF ASPERGILLUS

Two filamentous fungi commonly involved in ocular or periocular infections (and even more commonly tested) are *Aspergillus* and *Mucor*. *Aspergillus* is an acute branching (45 degree branching), septate fungus with parallel walls and fruiting bodies. *Mucor* is a right-angle branching (90 degree branching), non-septate fungus with nonparallel walls and broad hyphae.

The "**ASP**" in **ASP**-ergillus describes the fungus perfectly.

Acute branching (45 degree branching vs. *Mucor's* 90 degree branching)
Septate (vs. *Mucor* which is non-septate)
Parallel walls (vs. *Mucor's* nonparallel walls with broad hyphae)

Mnemonic origin unknown. Special thanks to R. Jean Campbell, MD for mentioning it.

JUNCTIONAL NEVUS OF THE CONJUNCTIVA IN ADULTS	Beware of any pathology report that diagnoses a junctional nevus of the conjunctiva in an adult. The correct diagnosis is likely primary acquired melanosis (PAM) of the conjunctiva which is a pre-malignant lesion.

Remember - A **JUNK**tional nevus of the con**JUNK**tiva is a **JUNK**y diagnosis in adults. The real diagnosis is probably PAM.

MELANOMA AND THE NEVUS OF OTA	The nevus of Ota is a congenital oculodermal melanosis involving the conjunctiva and eyelids. There is NO increased predisposition for conjunctival melanoma with nevus of Ota. However, there IS an increased risk of uveal, orbital, or meningeal melanoma.

In conjunctival nevus of **Ot**a, melanoma occurs in **Ot**a places.

That is to say, in conjunctival nevus of **Ot**a there is no increased risk of melanoma of the conjunctiva. The **Ot**a ("other") places where one can develop melanoma are the uvea, orbit, and meninges.

CONJUNCTIVAL BIOPSIES WITH SUSPECTED CRYSTALS	When you spot crystals in the conjunctiva, send the biopsy specimen in alcohol rather than formalin. Formalin will dissolve the crystals, but alcohol will preserve them for pathologic examination.

Just as you use fine **Crystal** to drink **Alcohol**, send suspected conjunctival **Crystals** to pathology preserved in **Alcohol** (not formalin).

CORNEA: EPITHELIAL EROSION VERSUS ULCERATION	In a corneal epithelial erosion, Bowman's membrane remains intact. However, in a corneal ulceration, Bowman's membrane is broken (violated).

Think of holding a **Bow** with its **bowstring** loosely down at your side. When the string of the **Bow** is **intact** (signifying an **intact Bowman's** layer), the string of the Bow is parallel to the ground. The Bow and its intact bowstring form the top of the lower-case letter "**e**," reminding you that in "**e**"-rosions Bowman's layer is intact. (See Illustration)

If the string of the **Bow** is **busted** (signifying a **broken Bowman's** layer), you flip the Bow upside-down to see what happened and repair the broken bowstring. The Bow now forms the lower-case letter "**u**," reminding you that in "**u**"-lceration Bowman's layer is broken. (See Illustration)

DESCEMET'S MEMBRANE: ULTRASTRUCTURAL COMPONENTS	Descemet's membrane can be divided into anterior and posterior portions. The anterior 1/3 of Descemet's is banded and is called the fetal layer because it develops in utero. The posterior 2/3 of Descemet's is non-banded and is formed after birth.

Think of a rock **Band** of **Ants** giving a concert on **Ant**erior surface of Descemet's membrane. With the **Band** of **Ants** on the **Ant**erior portion of Descemet's, the **Ant**erior portion is now is "**Banded**."

STAINS FOR MACULAR DYSTROPHY	Alcian blue stain and colloidal iron stain can both be used to stain for macular dystrophy. From the "Marilyn Monroe mnemonic" (see the Anterior Segment chapter) you know Alcian blue is a good stain for macular dystrophy. To remember that colloidal iron fits with Alcian blue use this mnemonic:

mnemonic — Look at the similarities in spelling: "**AL**c**I**a**N** Blue" & "Colloi**d**AL IroN."

Another useful hint to help you group these two stains for macular dystrophy together in you brain is this. Every time you hear or see the words "Alcian blue," say to yourself "Alcian blue & colloidal iron." The two stains will become inextricably linked together in your mind.

Note: See the Anterior Segment chapter for the full "Marilyn Monroe mnemonic" for corneal stromal dystrophies.

STAPHYLOMA VERSUS ECTASIA	An ectasia is a bulging of ocular structures that is NOT lined by uveal tissue, while a staphyloma is a bulging of ocular structures that IS lined by uveal tissue.

mnemonic — Remember - Sta**P**hylomas are usually **P**igmented since they are lined by uveal tissue.

Ectasias, by exclusion, are not lined by uveal tissue.

ALTERNATIVE STAINS FOR LATTICE DYSTROPHY	From the "Marilyn Monroe mnemonic" you know that **L**attice dystrophy contains **A**myloid which stains with **C**ongo red (the "**L.A. C**ounty" part of the mnemonic). Other stains that can be used to reveal the amyloid are thioflavin T, crystal violet, and even PAS.

mnemonic — There are lot's of "**T**'s" in La**TT**ice Dystrophy and its stains: **T**hioflavin **T** & Crys**T**al viole**T**.

As far as the "PAS" stain for amyloid goes, try the inextricably linked method mentioned above for the stains "Congo red and PAS" for lattice dystrophy.

VACUOLIZATION OF IRIS PIGMENT EPITHELIUM	Vacuolization of the iris pigment epithelium is commonly associated with diabetes mellitus. Other causes include Menkes kinky hair syndrome (a real disease, involving impaired copper metabolism), mucopolysaccharidoses, multiple myeloma, and it can be normal in the neonatal period.

mnemonic — Remember – The 5 "**M**'s" of iris pig**M**ent epitheliu**M** vacuoles.

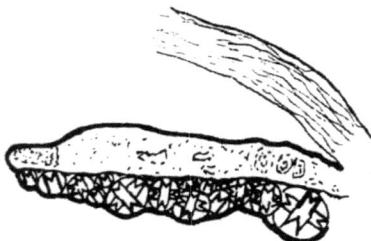

Imagine that under the microscope, lacy iris vacuolization is made up of many "**M**"-shaped doilies. (See Illustration)

Mellitus (diabetes **M**ellitus)
Menkes kinky hair syndrome
Mucopolysaccharidoses
Multiple **M**yeloma (only counts for 1 of the 5 "**M**'s")
Minors (normal in neonates)

Information source: Information source: Yanoff & Fine, Ocular Pathology: A Text and Atlas, 3rd ed. J.B. Lippincott, 1989, p. 565.

LACRIMAL GLAND TUMORS

Common lacrimal gland neoplastic tumors include lymphoproliferative lesions, epithelial neoplasms, pleomorphic adenomas, and carcinomas (including adenoid cystic carcinoma).

Half and half rule of lacrimal gland tumors.

1) Of the lacrimal gland neoplasms presenting without inflammatory signs or symptoms, approximately **half** are lymphoproliferative lesions, and **half** are epithelial neoplasms.

2) Of the epithelial neoplasms, approximately **half** are pleomorphic adenomas, and **half** are carcinomas.

3) Of the carcinomas approximately **half** are adenoid cystic tumors and **half** are others (including malignant mixed tumors, primary adenocarcinomas, mucoepidermoid carcinomas, and squamous carcinomas).

In order to use the **half and half** rule of lacrimal gland tumors, you must keep all of the tumors in the correct order: half lymphoproliferative -> half epithelial -> half pleomorphic adenoma -> half carcinomas -> half adenoid cystic carcinoma -> & half are others.

Step through the tumors in this order.

Of the non-inflammatory lacrimal gland tumors:
    half lymphoproliferative & half epithelial
        of epithelial:
    half pleomorphic adenoma & half carcinomas
        of carcinomas:
    half are adenoid cystic carcinoma & half are others

Information source: American Academy of Ophthalmology Basic and Clinical Science Course: Orbit, Eyelids, and Lacrimal System. American Academy of Ophthalmology, 2008, pp. 88-90.

RETAINED CELL NUCLEI IN LENS CELLS

The nuclei of the lens fibers are normally present only in the outer equatorial cortex. In certain conditions such as congenital rubella, Leigh's syndrome, Lowe's syndrome, and Trisomy 13 the nuclei may remain visible well into the nucleus of the lens.

Imagine that two retained cell nuclei are sitting in the lens talking to one another. One says to the other, "They **Tried 13** times to **Rub** me out, but I **Laid Low** and remained here in the nucleus."

Tried **13** times (**Tri**somy 13)
Rub me out (congenital **Rub**ella)
Laid (**Leigh's** syndrome)
Low (**Lowe's** syndrome)

Information source: Yanoff & Fine, Ocular Pathology: A Text and Atlas, 3rd edition. Lippincott, 1989, p. 38.

IMMUNE RESPONSE IN PHACOLYTIC GLAUCOMA

In phacolytic glaucoma liquefied lens cortex of a mature cataract leaks through an intact lens capsule into the aqueous and is engulfed by macrophages. This phenomenon causes a physical obstruction of the trabecular meshwork with no immune hypersensitivity reaction involved. On the other hand, phacoanaphylactic glaucoma is a sudden onset hypersensitivity reaction to lens proteins after cataract surgery or rupture of the lens capsule.

Remember – Phaco**L**ytic glaucoma **L**acks an immune hypersensitivity response.

Phaco**anaphylactic** glaucoma, by exclusion, is an immune hypersensitivity response; and **anaphylaxis** itself is a type of hypersensitivity – a dead giveaway!

HISTOLOGIC SUBTYPES OF ADENOID CYSTIC CARCINOMA

1) Adenoid cystic carcinoma of the lacrimal gland is a highly malignant epithelial tumor of the lacrimal gland with an overall ten year survival rate of only 20%. The histologic subtypes include cribriform (a "Swiss cheese" pattern with pseudo-ducts), comedocarcinoma (with central necrosis), cylindromatous (with tumor nests surrounded by thick basement membranes), basaloid (which is solid), sclerosing, and tubular (with true duct formation).

"**C**"**roy*** **BaST**e for subtypes of adenoid cystic carcinoma.

Imagine that your patient **roy** gets adenoid cystic carcinoma of the lacrimal gland. You know he has a poor prognosis and realize roy's "goose is cooked." Just like you would **BaST**e (baste) a goose before cooking it, thinking of poor Roy's fate makes you lament that you will soon, "**C**"**roy*** **BaST**e ("See Roy baste") because he has adenoid cystic carcinoma of the lacrimal gland. That is:

"**C**"**roy*** **BaST**e

Cribriform
Comedocarcinoma
Cylindromatous
Basaloid
Sclerosing
Tubular

*Note: The "**C**'s" of adenoid cystic carcinoma can be difficult to remember. Therefore, in this mnemonic the "**C**" in "**C**"**roy** stands for the first letter of the subtype and the "**roy**" stands for the second letter in the subtype.

2) The cribriform variant of adenoid cystic carcinoma is associated with a better prognosis for survival. The basaloid variant has the worst prognosis.

"**Cheese please**" but **B**asaloid is **B**ad for adenoid cystic prognosis.

"**Cheese please**" is a line from a famous commercial. You hope, for roy's sake, that the pathology turns out to be the cribriform (Swiss cheese) variant of adenoid cystic and think, "**Cheese please**." Basaloid has the worst prognosis, thus **B**asaloid is **B**ad.

CONFIGURATION OF ROSETTES IN RETINOBLASTOMA	Both Flexner-Wintersteiner rosettes and Homer Wright rosettes can be found in retinoblastoma. Flexner-Wintersteiner rosettes have a hollow central lumen while Homer Wright rosettes have no true lumen, since a tangle of neural filaments fills the central space in a Homer Wright rosette.

Flexner-Winter**Stein**er rosettes are like a beer **Stein** with a hollow center (a true lumen).

Homer Wright rosettes are like **Homer** Simpson who always finds himself **tangled** up in lots of trouble.

This reminds you that **Homer** Wright rosettes and **Homer** Simpson both have **tangles**.

VOGT-KOYANAGI-HARADA SYNDROME VERSUS SYMPATHETIC OPHTHALMIA	The choriocapillaris is involved in Vogt-Koyanagi-Harada syndrome (VKH), while the choriocapillaris is spared in sympathetic ophthalmia (SO).

SO stands for **S**paring **O**f the choriocapillaris,
and **VKH** stands for **V**ictimized **KH**oriocapillaris.

OR

Simply rationalize that VKH has more letters than SO, so more structures (including the choriocapillaris) must be involved in VKH.

Mnemonic by permission of and modified from Mary Mehaffey, MD.

THE EPONYMS ASSOCIATED WITH SCHWANNOMAS	Antoni A and Antoni B patterns are seen with schwannomas. The Antoni A pattern is characterized by areas of nuclear palisading, which are also called Verocay bodies. The Antoni B pattern denotes loose myxomatous areas.

First, Schw**AN**noma has **AN**toni patterns.

Antoni **A** is the palis-**A**-ding of nuclei in a Veroc-**A**-y body.

For Antoni **B**, rotate the **B** 90 degrees counterclockwise, and it resembles an **M**, reminding you of the **M**yxomatous areas of Antoni **B**.

PHASES OF MITOSIS	Mitosis is the process by which eukaryotic cells divide their nuclear material into two identical nuclei for the process of cell division. The four phases of mitosis are Prophase, Metaphase, Anaphase, and Telophase.

Microscopy "**Pro**s **M**eet **An**d **Tel**l" each other the phases of mitosis.

The four phases of mitosis are:
1. **Pro**s = **Pro**phase
2. **M**eet = **Met**aphase
3. **An**d = **An**aphase
4. **Tel**l = **Tel**ophase

CHOROIDAL MELANOMA: HISTOLOGIC PROGNOSIS

Spindle cell choroidal melanoma has a relatively good prognosis while the epithelioid variant has the worst prognosis.

mnemonic

Epitheli**oid** is the melanoma to **avoid** because is has the worst prognosis.

With **spindle** cell, you may live to be an old **spinster** because it has a good prognosis.

IRON INTRAOCULAR FOREIGN BODIES: THE FERROUS ION VERSUS THE FERRIC ION

Iron intraocular foreign bodies undergo decomposition leading to siderosis bulbi with damage to the retina, iris heterochromia, cataract and secondary glaucoma. The ferrous ion (Fe^{2+}) of iron is more toxic to the eye than the ferric ion (Fe^{+3}).

mnemonic

Ferr**ous** is more deleteri**ous** to the eye.

OR

mnemonic

Imagine **two** rotating **Ferris** wheels eating up an eyeball. (See Illustration)

The "**two**" reminds you that the Fe **two** plus ion (Fe^{2+}) which is called the Ferrous ("**Ferris**") ion causes more damage to the eye.

Information source: American Academy of Ophthalmology Basic and Clinical Science Course: Retina and Vitreous. American Academy of Ophthalmology, 2008, p. 326.

RHABDOMYOSARCOMA: HISTOLOGIC SUBTYPES

Rhabdomyosarcoma is the most common primary orbital malignancy in kids. Embryonal rhabdomyosarcoma is the most common histologic subtype and has the best prognosis (depending on site and metastasis). Botyroid rhabdomyosarcoma is a rare variant of embryonal. Pleomorphic (a.k.a. well differentiated) rhabdomyosarcoma may show striations of skeletal muscle differentiation. Alveolar rhabdomyosarcoma has the worst prognosis, and is usually located in the inferior orbit.

A **B**ranch of the **E.P.A.** (**E**nvironmental **P**rotection **A**gency) should be assigned to protect kids from rhabdomyosarcoma.

Botyroid
Embryonal (most common & best prognosis)
Pleomorphic
Alveolar (worst prognosis)

With **Alveolar**, **all** may be **over**,
It has an <u>Inferior</u> location with an <u>Inferior</u> (the worst) prognosis.

NOTES & YOUR OWN MNEMONICS:

NOTES & YOUR OWN MNEMONICS:

**SENSITIVITY &
SPECIFICITY**

1) Sensitivity and specificity are statistical measures of a test. For any particular test, the sensitivity is the number of patients who actually have a disease that are identified by the test as positive for the disease (the true positive test results divided by all the patients who actually have a disease). Conversely the specificity of a test is defined as the number of patients who are truly disease free that are identified by the test as negative for the disease (the true negative test results divided by all the patients who truly are disease free). So the sensitivity of a test is dependent upon the proportion of diseased patients in whom the test is positive. While the specificity of a test is dependent upon the proportion of healthy patients in whom the test is negative.

For **Sensitiv**ity, think of how **sensitive** patients are with **PID** (**P**elvic **I**nflammatory **D**isease).

That is to say, **Sensitiv**ity indicates the percentage of patients who test **P**ositive **I**n **D**isease (the number of patients identified by the test as positive for the disease out of all the patients who actually have the disease).

For **Specific**ity, think of the many **Specific**ations of the **NIH** (**N**ational **I**nstitute of **H**ealth).

That is to say, **Specific**ity indicates the percentage of patients who test **N**egative **I**n **H**ealth (the number of patients identified by the test as healthy (negative for disease) out of all the patients who are truly healthy and disease free).

2) Additionally, the sensitivity of a test gives an indirect indication of the false negative rate, while the specificity of a test gives and indirect indication of the false positive rate.

Se**N**sitivity gives an indirect indication of the false **N**egative rate.*

Se**N**sitivity has an "**N**" in it and gives an indirect indication of the false **N**egative rate. For example, a test with a sensitivity of 85% has a 15% false negative rate.

S**P**ecificity gives an indirect indication of the number of false **P**ositives.*

S**P**ecificity has a "**P**" and gives an indirect indication of the false **P**ositive rate. For example, a test with a specificity of 90% has a 10% false positive rate.

*Mnemonic by permission of and modified from Elliot Werner, MD.

ANTICHOLINERGIC TOXICITY

Atropine and other anticholinergic drugs commonly used in ophthalmology can lead to anticholinergic toxicity with some or all of the following signs mydriasis with cycloplegia, cutaneous flushing, hyperthermia, anhydrosis, CNS symptoms including agitation, delirium, lethargy, anxiety and even seizures, as well as urinary retention. Remember these signs of anticholinergic toxicity with this classic mnemonic:

> **Blind as a bat** and **Red as a beet**,
> **Hot as a hare** but **Dry as a bone**,
> & **Mad as a hatter** with **A full bladder**.

That is:
Blind as a bat (mydriasis with cycloplegia) and
Red as a beet (cutaneous flushing)
Hot as a hare (hyperthermia) but
Dry as a bone (anhydrosis)
Mad as a hatter (agitation, delirium, lethargy and/or anxiety) with
A full bladder (urinary retention)

Mnemonic origin unknown. Thank you to Michael Myles MD, FRCS(C) for mentioning it.

ESTER VERSUS AMIDE ANESTHETICS

In ophthalmology, the ester anesthetics, which include proparacaine and tetracaine, are typically administered topically. The amide anesthetics, which include lidocaine, bupivacaine, and mepivicaine, are typically administered in retrobulbar and peribulbar injections.

> **E**sters with an "**E**" are usually given topicall-"**E**."
> While **A**mides with an "**A**" are usually given retrobulb**A**r/peribulb**A**r.

Ester anesthetics given topically include proparacaine and tetracaine. Amides given retrobulbar and peribulbar include lidocaine, bupivacaine, and mepivicaine.

POSITIONING FOR SKULL X-RAYS

Classically skull x-rays were the imaging modality of choice for the bony structures of the orbit. While plain films have been largely replaced with CT imaging, patient positioning for skull views is of historical interest.

1) In the Waters view the head is extended (chin tilted up) so that the canthomeatal line is 37 degrees to the central beam. The Water's view is used for blow-out fractures.

> For the **Waters** view, tilt your chin up to "**sip water**."

> If you can't recall that the Water's view is used for blow-out fractures, imagine that **water** can extinguish a flame or you can simply **blow** it out (**Waters** view for a **blow** out fracture).

2) In the Caldwell view there is a 15 degree angle caudad tilt (forehead tilted down) to the posteroanteriorly directed central x-ray beam.

> For the **Caldwell** view, rest your forehead on a "**cold wall**."

MRI WEIGHTING AND THE EYE

1) When evaluating an MRI image, the weighting (T1 vs. T2) can be easily identified by looking at the vitreous cavity. On T2 images the vitreous appears bright, while on T1 images the vitreous is dark.

mnemonic On T2 MRI, H_2O in the vitreous makes it bright.

OR

mnemonic T2 MRI lights up the **2** eyes.

2) The term "T2" refers to the spin-spin relaxation time. The term "T1" refers to the spin-lattice relaxation time.

mnemonic The term "T**2**" is for **2 spins** (the **spin-spin** relaxation time).
The term "T**1**" is for the spin-"**1**"attice relaxation time
(use the numeral "**1**" to replace the "**l**" in the word "lattice").

THE PRE-OPERATIVE DISCUSSION

Before any ophthalmic surgical procedure, an adequate pre-operative discussion is a must. In terms the patient can understand, describe and explain the procedure, as well as the risks, benefits, and alternatives to the procedure. Then ask if the patient has any questions and answer them.

mnemonic For an adequate pre-operative discussion:
PARQ ("Park") yourself in front of the patient until everything has been discussed.

Procedure (explain the procedure)
Alternatives
Risk & benefits
Questions

Mnemonic by permission of and modified from Roger A. Dailey, Upper Eyelid Blepharoplasty, Focal Points, Vol 8, No 6, 1995.

SUCCINCT HISTORY FOR EMERGENCIES

When confronted with an ophthalmic emergency, the history may have to be abbreviated although it must be succinct and adequate.

mnemonic In an emergency, take an **AMPLE** history.

Allergies
Medications
Past history
Last meal
Events leading to presentation

Mnemonic origin unknown. Thank you to Mike Myles, MD, FRCS(C), for mentioning it.

THE BUSINESS OF EYE CARE

In addition to clinical knowledge and competence, consider the following factors when setting up and managing a successful ophthalmology practice as a successful business venture:

The **M**'s of Success in Eye Care.

Management
Managed care
Marketing
Money
Margins
Measurement
Multiples
Mission

Mnemonic by permission of and modified from R. Bruce Grene, MD and Richard L. Lindstrom, MD.

AND

Remember the **C**'s of being an Eye M.D. Be. . .

Competent
Cost-effective
Convenient
Caring

Mnemonic reprinted with permission from EyeNet Magazine, copyright 1998. Article by Elliot Finkelstein, EyeNet, September 1998, Vol 2, No. 9, p 5.

TREATMENT OF MALIGNANT HYPERTHERMIA

Malignant hyperthermia is a serious and potentially deadly complication of general anesthesia that must be treated promptly. Remember the treatment of malignant hyperthermia with this little rhyme:

Stop gas, Hypervent and **Dantroline**
Glu-In' Cold Fluids Bi the **Valentine.**

Stop gas – Stop triggering agent
Hypervent – Hyperventilate with a clean machine
Dantroline – Sounds like Dantrolene
*Glu – Glucose (sounds like "glue." Imagine you are gluing in a cold bag of intravenous fluid by the patient's heart)
In' – Insulin
Cold – Cool patient (cold i.v. solution, cooling blanket)
Fluids – Intravenous fluids, furosemide
Bi – Bicarbonate (sounds like "by")
Valentine – Monitor the heart & watch for ventricular tachycardia

Mnemonic by permission of and modified from "Stop Hot Dude Better Give Iced Fluids Fast," Aaron L. Zuckerberg, A Hot Mnemonic for the Treatment of Malignant Hyperthermia, Anesth Analg, 1993, Vol 77, p. 1077.

PATHOPHYSIOLOGY OF MALIGNANT HYPERTHERMIA	Most cases of malignant hyperthermia are caused by a mutation in the gene coding for ryanodine receptor 1 (RYR1). The ryadonine receptor protein is an ion channel that adjusts calcium concentrations, thereby regulating muscle contractions. In malignant hyperthermia, the channel is opened and calcium is released uncontrollably. The rising calcium concentration in the cell markedly increases the rate of metabolism, and the muscle contracts. (Mutations in the CACNA1S gene may also cause malignant hyperthermia by activating RYR1.) Remember the pathophysiology and mutation responsible for malignant hyperthermia with this mnemonic.

Ryan OD's on calcium and vomits calcium all over.

Ryan OD's sounds like ryanodine – the receptor that is mutated in malignant hyperthermia.
OD is an overdose
OD's on calcium (there is a high concentration of calcium in the cells)
and vomits calcium all over (releases calcium uncontrollably)

MALIGNANT HYPERTHERMIA AND DANTROLENE	Dantrolene is the most important intervention in the treatment of malignant hyperthermia.

Dan's the main man in the M. H. treatment plan.

Dan = dantrolene   **M. H.** – Malignant Hyperthermia

AND

Dantrolene sounds like "**Dan throw's line**."

Think that Dan throws you a life line when a patient has malignant hyperthermia.

AND

Malignant hyperthermia can be thought of as a struggle between two people called "**Ryan** and **Dan**". **Ryan** overdoses on calcium (the ryanodine receptor 1). Dan tries to save him by throwing him a safety line (dantrolene).

NOTES & YOUR OWN MNEMONICS:

NOTES & YOUR OWN MNEMONICS:

APPENDIX I. SUGGESTIONS FOR MULTIPLE-CHOICE EXAMS

1) The examiners who write multiple-choice exams probably have written or have used the same core textbooks you presently own to write the exam. Therefore when reading, imagine what questions you would ask if you were the examiner.

2) On the blank pages of your textbooks, (e.g. the AAO BCSC texts) write down the nitty-gritty details that you always have trouble remembering. Highlight important details in a special color such as pink. The night before the exam you will be able to quickly review these topics from the texts. If you're really short on time, just review the figures, photos, and tables in your textbooks.

3) Practice doing as many questions as you can, so that you'll have the mental stamina to sit there for several hours during the real exam. Lamkin's *Mass. Eye & Ear Infirmary Review Manual*, Chern's *Review Questions in Ophthalmology*, and the questions at the back of the AAO BCSC texts are helpful. The Osler course, Wills Eye Hospital course, and San Antonio course are some useful one-week intensive reviews that will help fill your brain with relevant facts.

4) If you're a resident taking the OKAP exam, see if a staff person or a board certified fellow will cover your call the night before the exam. If you're a fellowship candidate make sure all your patient commitments are handled the week before the exam. Arrange for the above coverage months ahead of time.

5) Read the questions carefully. Especially look for the words "NOT" or "EXCEPT" since they change the whole meaning of the question.

6) Read the answers carefully.

    i) If you know a response is inappropriate, cross it out (on the page if taking a paper test, mentally if taking a computerized test). That way, you'll have fewer options to distract you.

    ii) If there is only "one right answer," the manner in which the question is written often indicates the correct response. Ask yourself, "Why are they giving these particular answer choices?" If one answer is very similar to another answer or its converse, there is a higher chance that the correct response is one of the paired answers.

    iii) Qualifiers such as "never" and "always" usually suggest the statement is false.

    iv) Qualifiers such as "may" usually suggest the statement is true.

    v) An unusually long response is a clue that either the author is "snowing you" with gibberish, or that the author has to use a lot of words to ensure that the response is indeed correct.

7) For optics questions, don't spend more than a minute or two with each optics question. Remember the minus sign for diverging light rays. Watch the decimal place when working with millimeters, centimeters, and meters. Ask well beforehand whether calculators are allowed. If they aren't allowed make sure you remember your inverses from 1 to 12.

8) For the questions you have not answered, be sure to mark them distinctively so that you may return to them before turning in the completed exam. For the computerized ophthalmology board exam, there is an option to mark questions as "incomplete" so that you may return to them later. For written exams or the OKAP exam, put a distinctive mark on the question book itself (not the answer sheet) next to the questions you haven't answered so that later you can find them again.

9) For a written test with a computerized answer sheet and pencil, bring an eraser that doesn't smudge. Consider mechanical pencils that you won't have to sharpen. Ask well beforehand if you can bring snacks or a drink, if you are so inclined.

10) If using a computerized answer sheet and pencil, leave enough time to carefully transpose your answers to the computerized answer sheet.

11) To find more ophthalmology review material for written exams and ophthalmology CME programs in CD, DVD, and MP3 format, please visit www.ApolloAudiobooks.com and click on the "RESIDENTS & PHYSICIANS AUDIO" link. An excellent list of recommended review textbooks for written ophthalmology exams can also be found at www.ApolloAudiobooks.com in the "Ophtho for Residents" section.

APPENDIX II. SUGGESTIONS FOR ORAL EXAMS

1) Practice oral exam questions by talking to previous candidates and doing mock orals with former attendings, with current colleagues, or at the Osler course. Start making a list of sample oral questions in residency with your fellow residents. Organize the questions according to subspecialty, and you will have something to study from when it's time to prepare for the "big quiz."

Use an atlas or a question-and-answer book to get started making your own questions.

   i) R. J. Morris, Medical Ophthalmology for Postgraduate Examinations, Churchill Livingstone, 1991.

   ii) M. Ruben, Diagnostic Picture Tests in Ophthalmology, Year Book Medical Publishers, 1987.

   iii) J. F. Vander, Ophthalmology Secrets, Mosby, 2007.

2) Practice speaking slowly and clearly. Try answering questions out loud while looking at yourself in the mirror. Tape or video your answers to the questions and listen to them, or answer questions in front of a mirror.

3) Get to the test center a day ahead of time. If flying, bring your suit & tie as carry-ons. Get a good night's sleep.

4) Dress well in a suit & tie or unrevealing dress. Don't wear anything tight. No running shoes.

5) Smile pleasantly. Treat your examiner with the utmost respect and courtesy. Pretend you are Jimmy Stewart. Don't act like Howard Stern.

6) Ask for a pencil and paper if it is allowed. Make sure you have all the information. Ask:

   Is there another slide to go with this one?
   Can I see a later fluorescein, or fluorescein of the other eye?
   Can I see the field, acuity, or color vision testing of the other eye?
   Can I see the other cuts on this neuro-imaging scan?

7) Answer the question that is being asked. For example, when asked for a differential diagnosis, don't start talking about treatment plans.

8) Count to three before answering any question. Speak slowly. Never blurt out the answer even if you know it cold. Always make the examiner think he is taxing your brain a bit.

   <u>If you don't know the answer but think it will come to you later:</u>
   Consider stalling by repeating the relevant points.
   Spend more time describing the photo.
   Ask for more history, or ask the examiner to repeat the question.

   <u>If you're sure you will not get the answer:</u>
   Say, "I'm blocking. Can we come back to this later?"
   (Don't waste time by pretending not to understand the question.)
   Remember the examiner often has a quota of questions to run through in a certain period of time, and it is to your benefit to answer as many questions as possible.

9) Be concise. Give the most likely diagnosis or provisional diagnosis first. Then list the alternatives and finally the rare birds.

10) Exclude a life-threatening illness with ocular manifestations, and exclude any potentially blinding disorder.

In Kids: Exclude retinoblastoma or rhabdomyosarcoma. Treat amblyopia and refractive errors prior to surgery.

For funny photopsias: Exclude cancer associated retinopathy.

For visual loss or diplopia in an older patient: Exclude temporal arteritis.

For new onset mass or inflammation: Exclude sebaceous cell carcinoma, metastases, and the like.

For blind eyes with opaque media: Exclude melanoma by doing an ultrasound.

11) Do NOT use abbreviations. Say, "medial longitudinal fasciculus" rather than "MLF."

12) Use these common treatment plans:

Perform further investigations, including exams of family members.

Consider observation with follow-up.

Use medical treatment prior to laser or surgical treatment.

For very difficult problems, consider a second opinion from a subspecialist. (NOTE: You can always refer a patient to a subspecialist, but ensure you can give the examiner your opinion of what the consultant's management plan will be.)

13) Don't argue with the examiner. However, if you still think your answer is right, support your stance with additional information in a pleasant, non-offensive manner. If the questions start getting very tough, relax. You may actually be doing quite well and have reached the "bonus round."

14) STEM QUESTIONS: Be careful of a long, multiple part question, since the answers you give to the initial questions determine your answers for the subsequent ones. I don't recall any long stems on the oral American Board exams. If you're going astray, a competent and fair examiner will redirect you before you reach the end of the stem.

15) To find more ophthalmology review material for oral exams and ophthalmology CME programs in CD, DVD, and MP3 format, please visit www.ApolloAudiobooks.com and click on the "RESIDENTS & PHYSICIANS AUDIO" link. An excellent list of recommended review textbooks for ophthalmology oral exams can also be found at www.ApolloAudiobooks.com in the "Ophtho for Residents" section.

INDEX